RECENT ADVANCES IN OPERATOR THEORY AND OPERATOR ALGEBRAS

Recent Advances in Operator Theory and Operator Algebras

Hari Bercovici
David Kerr
Elias Katsoulis
Dan Timotin

CRC Press
Taylor & Francis Group
Boca Raton London New York

CRC Press is an imprint of the
Taylor & Francis Group, an **informa** business

A CHAPMAN & HALL BOOK

CRC Press
Taylor & Francis Group
6000 Broken Sound Parkway NW, Suite 300
Boca Raton, FL 33487-2742

© 2018 by Taylor & Francis Group, LLC
CRC Press is an imprint of Taylor & Francis Group, an Informa business

No claim to original U.S. Government works

Printed on acid-free paper
Version Date: 20170629

International Standard Book Number-13: 978-1-1380-3021-3 (Hardback)

This book contains information obtained from authentic and highly regarded sources. Reasonable efforts have been made to publish reliable data and information, but the author and publisher cannot assume responsibility for the validity of all materials or the consequences of their use. The authors and publishers have attempted to trace the copyright holders of all material reproduced in this publication and apologize to copyright holders if permission to publish in this form has not been obtained. If any copyright material has not been acknowledged please write and let us know so we may rectify in any future reprint.

Except as permitted under U.S. Copyright Law, no part of this book may be reprinted, reproduced, transmitted, or utilized in any form by any electronic, mechanical, or other means, now known or hereafter invented, including photocopying, microfilming, and recording, or in any information storage or retrieval system, without written permission from the publishers.

For permission to photocopy or use material electronically from this work, please access www.copyright.com (http://www.copyright.com/) or contact the Copyright Clearance Center, Inc. (CCC), 222 Rosewood Drive, Danvers, MA 01923, 978-750-8400. CCC is a not-for-profit organization that provides licenses and registration for a variety of users. For organizations that have been granted a photocopy license by the CCC, a separate system of payment has been arranged.

Trademark Notice: Product or corporate names may be trademarks or registered trademarks, and are used only for identification and explanation without intent to infringe.

Visit the Taylor & Francis Web site at
http://www.taylorandfrancis.com

and the CRC Press Web site at
http://www.crcpress.com

Contents

Preface   ix

Contributors   xi

1 Operator theory and Schubert calculus   **1**

   Hari Bercovici

  1.1 Three questions in operator theory 2
     1.1.1 Sums of Hermitian matrices 2
     1.1.2 Products of matrices 5
     1.1.3 Jordan models 5
  1.2 Schubert calculus 7
  1.3 The Littlewood–Richardson rule 11
  1.4 Practical intersection theory 15
  1.5 Back to operators 17
     1.5.1 Sums of Hermitian matrices 17
     1.5.2 Products of matrices 20
     1.5.3 Jordan models 21

References   23

2 Non-selfadjoint operator algebras: dynamics, classification and C^*-envelopes   **27**

   E.G. Katsoulis

  2.1 Introduction 27
  2.2 Examples 29
     2.2.1 The semicrossed product $C_0(X) \times_\sigma \mathbb{Z}^+$ 29
     2.2.2 The classification problem for semicrossed products . . 32
     2.2.3 The tensor algebra of a graph $\mathcal{G}$ 35
  2.3 C^*-correspondences 37
     2.3.1 The gauge-invariance uniqueness theorems 44
  2.4 Adding tails to a C^*-correspondence 45
     2.4.1 The Muhly–Tomforde tail 47
     2.4.2 The tail for (A, A, α) 47
  2.5 The C^*-envelope of an operator algebra 51
     2.5.1 The C^*-envelope of an arbitrary operator algebra . . . 56
  2.6 Dynamics and classification of operator algebras 58

v

Contents

vi

	2.6.1	Piecewise conjugate multisystems	60
	2.6.2	The multivariable classification problem	60
2.7	Crossed products of operator algebras		65
2.8	Local maps and representation theory		72

References   **77**

3 An introduction to sofic entropy   83
David Kerr

3.1	Introduction	83
3.2	Internal and external approximation	85
3.3	Amenable measure entropy	89
3.4	Amenable topological entropy	95
3.5	Sofic measure entropy	98
3.6	Sofic topological entropy	104
3.7	Dualizing sofic measure entropy	107
3.8	Algebraic actions	109
3.9	Further developments	111

References   **113**

4 The solution of the Kadison–Singer problem   117
Dan Timotin

4.1	Introduction		118
4.2	The Kadison–Singer problem		119
	4.2.1	Pure states	119
	4.2.2	The Kadison–Singer conjecture	120
	4.2.3	The paving conjecture	121
4.3	Intermezzo: what we will do next and why		123
	4.3.1	General plan	123
	4.3.2	Sketch of the proof	123
4.4	Analytic functions and univariate polynomials		124
	4.4.1	Preliminaries	125
	4.4.2	Nice families	126
4.5	Several variables: real stable polynomials		129
	4.5.1	General facts	129
	4.5.2	The barrier function	131
4.6	Characteristic and mixed characteristic polynomials		135
	4.6.1	Mixed characteristic polynomial	135
	4.6.2	Decomposing in rank one matrices and the characteristic polynomial	136
4.7	Randomization		138
	4.7.1	Random matrices and determinants	138
	4.7.2	Probability and partitions	140
4.8	Proof of the paving conjecture		141

4.9 Final remarks 144

References 147

Index 149

Preface

This book contains four chapters on different topics, with each being based on a series of lectures delivered at the OTOA-2014 and notes painstakingly prepared by each of the four eminent mathematicians, Professors Hari Bercovici (Indiana University Bloomington, USA), Elias G. Katsoulis (East Carolina University, USA), David Kerr (Texas A&M University, USA), and Dan Timotin (Institute of Mathematics of the Romanian Academy, Romania). This second biennial workshop-cum-conference, titled Recent Advances in Operator Theory and Operator Algebras-2014 (OTOA-2014), was held at the Bangalore centre of the Indian Statistical Institute (ISI) during December 09–December 19, 2014 and attended by about a hundred participants from around the world.

We express our gratitude to the sponsors of OTOA-2014, Indian Statistical Institute, National Centre for Mathematics, and Jawaharlal Nehru Centre for Advanced Scientific Research, India, for their generous support. We also extend special thanks to all the staff of the guest house, the Statistics and Mathematics unit, canteen, and accounts and transport offices of ISI Bangalore centre for their support to OTOA-2014. For the final stage of formatting this book, our thanks go to Amit Maji.

Organizers of OTOA-2014

B. V. R. Bhat and Jaydeb Sarkar from Indian Statistical Institute, Bangalore, and V. S. Sunder from Institute of Mathematical Sciences, Chennai

Contributors

Hari Bercovici
Indiana University
Bloomington, IN, USA

E.G. Katsoulis
East Carolina University
Greenville, NC, USA

David Kerr
Texas A&M University
College Station, TX, USA

Dan Timotin
Institute of Mathematics of the
Romanian Academy
Bucharest, Romania

Chapter 1

Operator theory and Schubert calculus

Hari Bercovici

1.1	Three questions in operator theory	2
	1.1.1 Sums of Hermitian matrices	2
	1.1.2 Products of matrices	5
	1.1.3 Jordan models	5
1.2	Schubert calculus	7
1.3	The Littlewood–Richardson rule	11
1.4	Practical intersection theory	15
1.5	Back to operators	17
	1.5.1 Sums of Hermitian matrices	17
	1.5.2 Products of matrices	20
	1.5.3 Jordan models	21

Acknowledgment. The author was supported in part by grants from the National Science Foundation.

Abstract. Several quite distinct problems in operator theory concerning eigenvalues, singular values, and Jordan models have virtually identical answers. We explain how these answers can all be derived from Schubert calculus, that is, from the intersection theory of the Grassmann manifold. The connection between intersection theory and the characterization of the eigenvalues of a sum of Hermitian matrices was known for some time and was completed by the work of A. Klyachko, A. Knutson, and T. Tao. The connection with the other questions alluded to above was found more recently. We discuss finite- as well as infinite-dimensional versions of these problems, some of which are not completely resolved.

1.1 Three questions in operator theory

1.1.1 Sums of Hermitian matrices

We start with the best known of the three questions. Denote by M_n the algebra of $n \times n$ complex matrices. Given a Hermitian matrix $X \in M_n$, that is, $X = [x_{ij}]_{i,j=1}^n$ and $\overline{x_{ij}} = x_{ji}$, $i, j = 1, \ldots, n$, we denote by

$$\Lambda(X) = (\lambda_1(X) \geq \cdots \geq \lambda_n(X))$$

the eigenvalues of X, repeated according to multiplicity.

Problem. Suppose that $A, B \in M_n$ are Hermitian and $\Lambda(A)$, $\Lambda(B)$ are known. Characterize the possible values of $\Lambda(A + B)$.

Some of the earliest results about this problem were obtained by H. Weyl [26]. A system of inequalities characterizing the possible $\Lambda(A + B)$ was conjectured by A. Horn and verified for small values of n [14]. This came to be known as the Horn problem. The connection with intersection theory was observed by several mathematicians (see, for instance, [13, 15]). This connection yields a system of inequalities which $\Lambda(A + B)$ must necessarily satisfy. The sufficiency of these inequalities (along with the trace identity $\sum_{j=1}^n (\lambda_j(A) + \lambda_j(B) - \lambda_j(A + B)) = 0$) was established by Klyachko [19] who also noted that this proves the Horn conjecture provided a certain combinatorial fact—the saturation conjecture—is proved. The saturation conjecture was established soon after by Knutson and Tao [21] (see also [9]). This story is recounted by Fulton in his beautiful survey [12].

To see how projective geometry is related to this problem, we prove Weyl's inequality

$$\lambda_{i+j-1}(A + B) \leq \lambda_i(A) + \lambda_j(B), \tag{1.1}$$

provided that $i + j - 1 \leq n$. Chose orthonormal bases $\{e_i\}_{i=1}^n$, $\{f_i\}_{i=1}^n$, and $\{g_i\}_{i=1}^n$ in $\mathbb{C}^n$ such that

$$Ae_i = \lambda_i(A)e_i, \quad Bf_i = \lambda_i(B)f_i, \quad (A + B)g_i = \lambda_i(A + B)g_i, \quad i = 1, \ldots, n.$$

For any unit vector $x \in \mathbb{C}^n$ we have the identity

$$\langle Ax, x \rangle + \langle Bx, x \rangle = \langle Cx, x \rangle,$$

where $\langle u, v \rangle$ denotes the scalar product of u and v. The inequality (1.1) follows for an appropriate choice of x. Suppose, for instance, that x is in the linear space M spanned by $\{e_i, e_{i+1}, \ldots, e_n\}$. Then

$$\langle Ax, x \rangle \leq \lambda_i(A),$$

as can be seen by writing x in this basis. Similarly, when x belongs to the space N spanned by $\{f_j, f_{j+1}, \ldots, f_n\}$, we have

$$\langle Bx, x \rangle \leq \lambda_j(B).$$

Operator theory and Schubert calculus 3

On the other hand, if x belongs to the space Q spanned by $\{g_1, \ldots, g_{i+j-1}\}$, we have

$$\langle (A+B)x, x \rangle \geq \lambda_{i+j-1}(A+B).$$

Thus (1.1) follows from these three inequalities provided that the intersection $M \cap N \cap Q$ can be shown to contain nonzero vectors. This follows simply because

$$\dim(M) + \dim(N) + \dim(Q) = 2n + 1 > 2n.$$

More complicated inequalities can be deduced by noting that

$$\mathrm{Tr}(AP) + \mathrm{Tr}(BP) = \mathrm{Tr}((A+B)P)$$

for any orthogonal projection P and by choosing an appropriate projection. We need to explain the meaning of the word 'appropriate' in this context. Using the orthonormal bases introduced above, denote by G_k the linear space spanned by $\{g_1, \ldots, g_k\}$, $k = 1, \ldots, n$. Suppose that $K = \{k_1 < k_2 < \cdots < k_r\}$ is a subset of $\{1, \ldots, n\}$ and let P be the orthogonal projection onto a space M of dimension r such that

$$\dim(M \cap G_{k_x}) \geq x, \quad x = 1, \ldots, r. \tag{1.2}$$

Then the inequality

$$\mathrm{Tr}((A+B)P) \geq \sum_{x=1}^{r} \lambda_{k_x}(A+B)$$

holds. The verification is fairly straightforward, using only the fact that M has an orthonormal basis $\{h_1, \ldots, h_r\}$ with the property that $h_x \in M_{k_x}$ for $x = 1, \ldots, r$. Once this basis is chosen,

$$\mathrm{Tr}((A+B)P) = \sum_{x=1}^{r} \langle (A+B)h_x, h_x \rangle,$$

and $\langle (A+B)h_x, h_x \rangle \geq \lambda_{k_x}(A+B)$, $x = 1, \ldots, r$. In order to obtain upper estimates for $\mathrm{Tr}(AP)$ we use instead the linear space E_{n+1-i} spanned by $\{e_i, \ldots, e_n\}$. (The subscripts are chosen so $\dim(E_i) = i$.) When $I = \{i_1 < \cdots < i_r\}$ and the range M of P satisfies the inequalities

$$\dim(M \cap E_{n+1-i_x}) \geq r + 1 - x, \quad x = 1, \ldots, r, \tag{1.3}$$

we have $\mathrm{Tr}(AP) \leq \sum_{x=1}^{r} \lambda_{i_x}(A)$. Similarly, let F_{n+i-j} be the linear space spanned by $\{f_j, \ldots, f_n\}$, let $J = \{j_1 < \cdots < j_r\}$ be a subset of $\{1, \ldots, n\}$, and suppose that

$$\dim(M \cap F_{n+1-j_x}) \geq r + 1 - x, \quad x = 1, \ldots, r. \tag{1.4}$$

Then we have $\mathrm{Tr}(BP) \leq \sum_{x=1}^{r} \lambda_{j_x}(B)$. If the projection P can be chosen so all three sets of conditions (1,2–4) are satisfied by its range M, we deduce the inequality

$$\sum_{k \in K} \lambda_k(A + B) \leq \sum_{i \in I} \lambda_i(A) + \sum_{j \in J} \lambda_j(B).$$

The Horn problem has an easily described analog for compact operators on an infinite-dimensional Hilbert space $\mathcal{H}$. Suppose for simplicity that A is such an operator and $A \geq 0$, that is, $\langle Ah, h \rangle \geq 0$ for all $h \in \mathcal{H}$. In this case, the positive eigenvalues of A can be arranged in a sequence

$$\Lambda(A) = \{\lambda_1(A) \geq \lambda_2(A) \geq \cdots\},$$

where each eigenvalue is repeated according to its multiplicity. When A has infinite rank, this sequence consists of strictly positive terms and it tends to zero. When A has finite rank, we simply set $\lambda_j(A) = 0$ when j exceeds the rank of A. The analog of the Horn problem asks for a description of the set

$$\{(\Lambda(A), \Lambda(B), \Lambda(A + B)) : A, B \text{ compact and } \geq 0\}.$$

The problem can be formulated for arbitrary compact operators as well. In that case $\Lambda(A)$ must be replaced by two sequences converging to zero, one consisting of the positive and the other of the negative eigenvalues of A. See [3] for a precise formulation.

There is another infinite-dimensional analog of the Horn problem that involves elements of a factor of type II_1. A factor of type II_1 is simply an infinite-dimensional algebra $\mathcal{A}$ of bounded operators on a Hilbert space $\mathcal{H}$ with the following three properties.

1. $\mathcal{A}$ is closed in the weak operator topology.

2. $\mathcal{A}$ is simple, that is, it has no nontrivial two-sided ideals.

3. There exists a nonzero, positive linear functional $\tau : \mathcal{A} \to \mathbb{C}$ such that $\tau(ST) = \tau(TS)$ for every $S, T \in \mathcal{A}$.

A factor $\mathcal{A}$ of type II_1 always has a multiplicative unit 1 (which can be assumed to be the identity operator on $\mathcal{H}$) and the functional τ is often normalized by the condition $\tau(1) = 1$. The positivity condition means that $\tau(T^*T) \geq 0$, $T \in \mathcal{A}$. The normalized functional τ, called *the* trace on $\mathcal{A}$, is unique and it is also continuous when $\mathcal{A}$ is endowed with the weak operator topology.

Suppose now that $\mathcal{A}$ is a II_1 factor with trace τ and $T = T^* \in \mathcal{A}$. The spectral theorem for T implies the existence of a spectral measure e_T defined on the Borel subsets of $[0, 1]$ with values in $\mathcal{A}$, and the existence of a nonincreasing function $\lambda_T : [0, 1] \to \mathbb{R}$ such that

$$T = \int_0^1 \lambda_T(t)e(dt),$$

and $\tau(\sigma) = |\sigma|$ for every Borel set $\sigma \subset [0, 1]$. Here we use $|\sigma|$ to denote the

Operator theory and Schubert calculus 5

Lebesgue measure of σ. The function λ_T is a substitute for the eigenvalues of T and it is uniquely determined if we assume it to be right continuous on $[0, 1)$ and left continuous at 1. Of course, T only has actual eigenvalues when the function λ_T is constant on some open interval. The analog of the Horn problem asks for a characterization of the triples $(\lambda_S, \lambda_T, \lambda_{S+T})$, where S, T are selfadjoint elements of some II$_1$ factor. See [5] for a detailed discussion.

1.1.2 Products of matrices

Given a matrix $A \in M_n$, we denote by $|A| = (A^*A)^{1/2}$ its absolute value. The numbers in the sequence $\Lambda(|T|)$ are known as the *singular values* of T.

Problem. Suppose that $A, B \in M_n$ are Hermitian and $\Lambda(|A|), \Lambda(|B|)$ are known. Characterize the possible values of $\Lambda(|AB|)$.

This problem is discussed in [20] for invertible matrices A and B. See [4] for the general case. The solution of the problem is best expressed in terms of the numbers $\log \lambda_j(|A|), \log \lambda_j(|B|), \log \lambda_j(|A + B|)$. These sequences turn out to satisfy precisely the same inequalities as $\Lambda(A), \Lambda(B)$, and $\Lambda(A + B)$ when A and B are Hermitian. We show later how to connect this directly to the intersection theory.

The II$_1$ version of this problem is easily stated. One asks for a characterization of the triples $(\lambda_{|S|}, \lambda_{|T|}, \lambda_{|ST|})$ when S and T are elements of some II$_1$ factor. This problem is discussed in [4].

Another infinite-dimensional version of this problem can be formulated for operators T on an infinite-dimensional Hilbert space $\mathcal{H}$ under the assumption that $1 - T$ is a compact operator. In this case, the singular values of T form a sequence of positive numbers that converges to 1. The problem is the subject of a forthcoming work.

1.1.3 Jordan models

Given an integer $k \geq 0$, denote by J_k the Jordan cell of size k. Thus, J_k is an operator acting on the standard basis $\{e_1, \ldots, e_k\}$ of $\mathbb{C}^k$ as follows:

$$J_k e_1 = 0, J_k e_{i+1} = e_i, \quad i = 1, \ldots, k - 1.$$

The Jordan model theorem tells us that every nilpotent matrix T is similar to a matrix of the form $J_{k_1} \oplus J_{k_2} \oplus \cdots \oplus J_{k_n}$, where $k_1 \geq \cdots \geq k_n$. The numbers k_i are uniquely determined by T, and we write $\mu(T) = \{k_1, \ldots, k_n\}$ for the collection of these integers. For ease of notation, it is convenient to identify $\{k_1, \ldots, k_n\}$ with $\{k_1, \ldots, k_n, 0\}$. In other words, we allow an arbitrary number of Jordan cells of size zero.

Suppose now that T is a nilpotent matrix, M is an invariant subspace for T, and set $T' = T|M$ and $T'' = P_{M^\perp}T|M^\perp$.

Problem. Characterize the collection of triples $(\mu(T'), \mu(T''), \mu(T))$.

6 Recent Advances in Operator Theory and Operator Algebras

This is a problem of a rather different nature from the first two. In particular, the vector $\mu(T)$ consists of nonnegative integers. Nonetheless, the answer to this question is given by the same set of inequalities. We show later how this answer is again explained by the intersection theory.

The preceding problem was first considered by Klein [17] in the somewhat more general context of finitely generated torsion modules over a discrete valuation ring. (A discrete valuation ring is simply a principal ideal domain with a unique prime ideal. The relevant ring for the nilpotent problem is the ring $\mathbb{C}[[x]]$ of formal power series in one variable.) A somewhat different proof can be found in [25]. The result expresses the solution of the problem in terms of the combinatorial Littlewood–Richardson rule which we review later. The connection with the intersection theory is made in [6].

We review one of the possible infinite-dimensional versions of Problem 1.1.3 involving operator semigroups. Suppose that $\{T(t) : t \geq 0\}$ is a strongly continuous semigroup of operators on a Hilbert space $\mathcal{H}$. In other words, each $T(t)$ is a bounded operator on $\mathcal{H}$, $T(0) = 1$, $T(t+s) = T(t)T(s)$ for all $t, s \geq 0$, and the map $t \mapsto T(t)h$ is continuous for every $h \in \mathcal{H}$. Arbitrary semigroups are difficult to classify up to unitary equivalence and even up to similarity. There is a class of semigroups which can be classified up to *quasisimilarity*, whose definition we now recall. Two semigroups $\{T(t) : t \geq 0\}$ on $\mathcal{H}$ and $\{T'(t) : t \geq 0\}$ on $\mathcal{H}'$ are said to be quasisimilar if there exist bounded linear operators $X : \mathcal{H} \to \mathcal{H}'$ and $Y : \mathcal{H}' \to \mathcal{H}$ that are one-to-one and have dense ranges such that

$$XT(t) = T'(t)X, \quad T(t)Y = YT'(t), \quad t \geq 0.$$

The semigroups that we can classify are the nilpotent semigroups, defined by the requirement that $T(t) = 0$ for some $t > 0$ (and hence $T(s) = 0$ for all $s \geq t$). The building blocks for this class are defined as follows. Given a positive number α, define $\mathcal{H}_\alpha$ to be the space $L^2((0, \alpha))$ (relative to Lebesgue measure) and define $\{J_\alpha(t) : t \geq 0\}$ by setting

$$(J_\alpha(t)f)(x) = \begin{cases} f(x - t), & x \leq t < \alpha, \\ 0, & \text{otherwise.} \end{cases}$$

Clearly $J_\alpha(\alpha) = 0$ but $J_\alpha(t) \neq 0$ for $t \in (0, \alpha)$. The classification theorem (see [2, 10]) shows that every nilpotent semigroup $T = \{T(t) : t \geq 0\}$, acting on a separable Hilbert space, is quasisimilar to a uniquely determined semigroup of the form

$$J_{\alpha_1}(t) \oplus J_{\alpha_2}(t) \oplus \cdots, \quad t \geq 0,$$

where $\alpha = (\alpha_1, \alpha_2, \dots)$ is a nonincreasing sequence of nonnegative numbers. We set $\mu(T) = \alpha$.

Suppose now that the semigroup T has an invariant subspace M, that is, $T(t)M \subset M$ for every $t \geq 0$. Then the restricted semigroup T' defined by $T'(t) = T(t)|M$, $t \geq 0$, is again a nilpotent semigroup. Similarly, one obtains

Operator theory and Schubert calculus

another nilpotent semigroup T'' by compressing eatch $T(t)$ to $\mathcal{M}^{\perp}$. One can then ask about the relations that $\mu(T), \mu(T')$, and $\mu(T'')$ must satisfy. The connection with intersection theory is made in [7] in a more general context. The solution is complete when the sequence $\mu(T)$ converges to zero, but not for arbitrary $\mu(T)$. More precisely, the inequalities which arise from intersection theory are still necessary in the general case but they are not known to be sufficient.

1.2   Schubert calculus

Hermann Schubert was an early master of enumerative geometry. He solved numerous counting problems, of which we show one example below. His method often consisted of looking at a carefully chosen special case of the problem and then concluding, often without rigorous justification, that the result obtained in the special case is generically the correct result. Hilbert's 15th problem [16] asks for establishing a rigorous foundation for Schubert's calculations and for the creation of effective methods for the calculation of intersection numbers. The first goal was achieved by the creation of various cohomology and intersection theories where Schubert's numbers can be calculated algebraically, at least in principle; see [11] for a detailed technical exposition. The calculations can be quite difficult in practice. At the time [16] was written it could not be ascertained whether Schubert's answer to the following problem is correct: Given five quadrics in three-dimensional space, how many quadrics are tangent to all five? (Answer: 666,841,048.) The following problem will serve to illustrate Schubert's method. Given four lines in a three-dimensional space, how many lines intersect all four? Schubert starts by moving the four lines to a special position in which two of them intersect in a point A and determine a plane P and the other two intersect in a point B and determine a plane Q. It is easy to see that there are two solutions: the line joining A and B, and the intersection of P with Q. Schubert concludes by the "principle of continuity" that the general answer is 2.

Even in the special problem just discussed, it is quite clear that some adjustments must be made to obtain the correct answer. For instance, the planes P and Q could be parallel so it is better to consider the problem in projective space that allows for lines at infinity. In general, finding the two points amounts to solving a quadratic equation. Not all quadratic equations with real coefficients have real roots, so it is also preferable to work with complex rather than real three-dimensional space.

We give a brief description of the intersection theory that is relevant to the problems stated in Section 1.1. Given positive integers r, n such that $r < n$, we denote by $G(n, r, \mathbb{C})$ the collection of all subspaces $M \subset \mathbb{C}^n$ of dimension

8   *Recent Advances in Operator Theory and Operator Algebras*

r. A natural topology is introduced on $G(n, r, \mathbb{C})$ by the metric

$$d(M, N) = \|P_M - P_N\|,$$

where $\|T\|$ denotes the usual operator norm of T, and P_M is the orthogonal projection with range M. Endowed with this topology, $G(n, r, \mathbb{C})$ is a complex manifold of (complex) dimension $r(n - r)$. It is usually called a *Grassmann manifold*. In order to understand the topology of this manifold it is useful to see it as a union of *cells*, each one homeomorphic to $\mathbb{C}^k$ for some k. This is done as follows. Fix a complete flag $\mathcal{E} = (E_j)_{j=1}^{n}$ in $\mathbb{C}^n$, that is, $E_j \subset \mathbb{C}^n$ is a linear space of dimension j, and

$$E_0 \subset E_1 \subset \cdots \subset E_n.$$

For any $M \in G(n, r, \mathbb{C})$, the sequence of integers

$$\dim(M \cap E_i), \quad i = 0, \ldots, n,$$

increases from 0 to r and the difference between successive terms is 0 or 1. It follows that there are uniquely determined integers $0 \le i_1(M) < \cdots < i_r(M) \le n$ such that

$$\dim(M \cap E_{i_x(M)}) = x = \dim(M \cap E_{i_x(M)-1}) + 1, \quad x = 1, \ldots, r.$$

Given a set $I = \{i_1 < \cdots < i_r\} \subset \{1, \ldots, n\}$, we denote by $s(\mathcal{E}, I)$ the collection of those spaces $M \in G(n, r, \mathbb{C})$ with the property that $i_x(M) = i_x$, $x = 1, \ldots, r$. The set $s(\mathcal{E}, I)$ is called the *Schubert cell* determined by the flag $\mathcal{E}$ and the set I. The Schubert cell $s(\mathcal{E}, I)$ is homeomorphic to $\mathbb{C}^{d_I}$, where

$$d_I = \sum_{x=1}^{r}(i_x - x).$$

This is seen as follows. Fix vectors $e_1, \ldots, e_n$ such that E_i is spanned by $\{e_1, \ldots, e_i\}$ for $1 = 1, \ldots, n$. A space $M \in s(\mathcal{E}, I)$ has a basis $m_1, \ldots, m_r$ such that

$$m_x = \sum_{i=1}^{i_x} a_{ix} e_i, \quad x = 1, \ldots, r.$$

This basis is uniquely determined if we require the coefficients to satisfy

$$a_{ix} = \begin{cases} 1, & i = i_x, \\ 0, & i = i_y \text{ for some } y < x. \end{cases}$$

One can then parameterize $s(\mathcal{E}, I)$ by the undetermined coefficients a_{ix}, of which there are d_I. The *Schubert variety* $\mathfrak{S}(\mathcal{E}, I)$ is the closure of $s(\mathcal{E}, I)$ in $G(n, r, \mathbb{C})$ and it can be described as

$$\mathfrak{S}(\mathcal{E}, I) = \{M \in G(n, r, \mathbb{C}) : \dim(M \cap E_{i_x}) \ge x, x = 1, \ldots, r\}.$$

It is easily verified that the boundary of $\mathfrak{S}(\mathcal{E}, I)$ is contained in a union of Schubert cells of dimension less than d_I. Elementary algebraic topology implies that the integral homology $H_*(G(n, r, \mathbb{C}))$ of $G(n, r, \mathbb{C})$ has a set of generators $\{\sigma_I\}_I$ indexed by the sets $I \subset \{1, \ldots, n\}$ of cardinality r, one generator σ_I corresponding to each Schubert variety $\mathfrak{S}(\mathcal{E}, I)$. For instance, the class σ_I corresponds to one point in $G(n, r, \mathbb{C})$ if $I = \{1, \ldots, r\}$, and σ_I is the class of the entire manifold $G(n, r, \mathbb{C})$ if $I = \{n - r + 1, \ldots, n\}$. Intersection theory endows $H_*(G(n, r, \mathbb{C}))$ with an associative ring structure. We denote the product by the symbol $\cap$. The unit is $1 = \sigma_{\{n-r+1,\ldots,n\}}$ and we also denote by $\bullet = \sigma_{\{1,\ldots,r\}}$ the class of one point. An equality of the form $\sigma_I \cap \sigma_J = 5\bullet$ has the following interpretation in terms of actual intersections of Schubert varieties: given flags $\mathcal{E}$ and $\mathcal{F}$, the intersection $\mathfrak{S}(\mathcal{E}, I) \cap \mathfrak{S}(\mathcal{F}, J)$ is always nonempty. For *generic* flags, this intersection contains exactly five distinct points, and these points are in $s(\mathcal{E}, I) \cap s(\mathcal{F}, J)$. Genericity can be interpreted as belonging to a dense set, where density can be taken in the usual gap metric of (sequences of) spaces, or in the sense of Zariski.

The four-line problem formulated above is easily cast in the formal framework of the intersection ring. First, a point in $\mathbb{C}^3$ corresponds to a linear subspace of dimension 1 in $\mathbb{C}^4$, and a line in $\mathbb{C}^3$ corresponds to a linear subspace of dimension 2 in $\mathbb{C}^4$. Thus the question pertains to the Grassman manifold $G(4, 2, \mathbb{C})$. Two elements $M, N \in G(4, 2, \mathbb{C})$ correspond to intersecting lines in $\mathbb{C}^3$ if $\dim(M \cap N) \geq 1$. Equivalently, $M \in \mathfrak{S}(\mathcal{E}, I)$, where $\mathcal{E}$ is a flag satisfying $E_2 = N$ and $I = \{2, 4\}$. The statement that there are two lines intersecting four given lines becomes simply

$$\sigma_{\{2,4\}}^4 = \sigma_{\{2,4\}} \cap \sigma_{\{2,4\}} \cap \sigma_{\{2,4\}} \cap \sigma_{\{2,4\}} = 2\bullet,$$

and this can be verified if we know how to multiply the generators σ_I of $H_*(G(4, 2, \mathbb{C}))$.

Returning to $H_*(G(n, r, \mathbb{C}))$ for general n and r, the multiplicative structure is determined by knowledge of the constants c_{IJ}^K in the equality

$$\sigma_I \cap \sigma_J = \sum_K c_{IJ}^K \sigma_K, \tag{1.1}$$

where I, J, and K are subsets with cardinality r of $\{1, \ldots, n\}$, and $c_{IJ}^K = 0$ unless $d_K = d_I + d_J - r(n - r)$. These constants are nonnegative integers called the *Littlewood–Richardson coefficients*. Littlewood and Richardson were the first to state a combinatorial method (the Littlewood–Richardson rule) for calculating these numbers. The first (incomplete) proof appears in [24] and it is rumored [25, p.148] that the rule found concrete applications in the Appollo space program before a complete proof was found. A simple recent proof is given in [23] in the original context of irreducible representations of the general linear group.

We argue first that it suffices to calculate the products $\sigma_I \cap \sigma_J \cap \sigma_K = c_{IJK}\bullet$ in case $d_I + d_J + d_K = 2r(n - r)$. This last condition means that the sum of the

10   *Recent Advances in Operator Theory and Operator Algebras*

codimensions of σ_I, σ_J, and σ_K is exactly equal to the dimension of $G(n, r, \mathbb{C})$, thus ensuring that the intersection is generically a discrete collection of points. To do this we introduce for each set $I = \{i_1 < \cdots < i_r\} \subset \{1, \ldots, n\}$ the set

$$\widetilde{I} = \{n + 1 - i_r < \cdots < n + 1 - i_1\}.$$

The generator $\widetilde{\sigma}_I = \sigma_{\widetilde{I}}$ is called the *Poincaré dual* of σ_I.

Lemma 1.2.1. *Suppose that $I, J \subset \{1, \ldots, n\}$ are subsets of cardinality r such that $d_I + d_J = r(n - r)$. Then*

$$\sigma_I \cap \sigma_J = \begin{cases} \bullet, & J = \widetilde{I} \\ 0, & \text{otherwise.} \end{cases}$$

Proof. Let $\mathcal{E}$ and $\mathcal{F}$ be two arbitrary flags in $\mathbb{C}^n$. Suppose first that $J = \widetilde{I}$, so a space $M \in \mathfrak{S}(\mathcal{E}, I) \cap \mathfrak{S}(\mathcal{F}, J)$ must satisfy

$$\dim(M \cap E_{i_x}) \geq x, \quad \dim(M \cap F_{n+1-i_x}) \geq r + 1 - x, \quad x = 1, \ldots, r.$$

Observe that the spaces E_{i_x} and F_{n+1-i_x} most have nonzero intersection, generically of dimension 1. Chose a vector v_x in this intersection. Generically, the vectors $v_1, \ldots, v_r$ are linearly independent and generate a space in $\mathfrak{S}(\mathcal{E}, I) \cap \mathfrak{S}(\mathcal{F}, J)$.

Suppose now that $J \neq \widetilde{I}$. In this case we have $i_x + j_{r+1-x} \leq n$ for some $x = 1, \ldots, r$, and therefore generically $E_{i_x} \cap F_{j_{r+i-x}} = \{0\}$. On the other hand, for every $M \in \mathfrak{S}(\mathcal{E}, I) \cap \mathfrak{S}(\mathcal{F}, J)$ we have

$$\dim(M \cap E_{i_x}) + \dim(M \cap F_{j_{r+i-x}}) = x + r + 1 - x > r,$$

and thus the two subspaces $M \cap E_{i_x}$ and $M \cap F_{j_{r+1-x}}$ of M must have nontrivial intersection, in contradiction with $E_{i_x} \cap F_{j_{r+i-x}} = \{0\}$. Thus, generically, no such spaces M exist. $\qquad\square$

Corollary 1.2.2. *Suppose that I, J and K are subsets of cardinality r of $\{1, \ldots, n\}$ such that $d_I + d_J + d_K = 2r(n - r)$. Then*

$$\sigma_I \cap \sigma_J \cap \sigma_K = c_{IJ}^{\widetilde{K}} \bullet .$$

In other words, $c_{IJK} = c_{IJ}^{\widetilde{K}}$.

Proof. Multiply the equality

$$\sigma_I \cap \sigma_J = \sum_L c_{IJ}^K \sigma_L$$

by σ_K and apply Lemma 1.2.1 to see that the only term remaining on the right-hand side corresponds to $L = \widetilde{K}$. $\qquad\square$

1.3 The Littlewood–Richardson rule

The coefficients c_{IJK} discussed in Section 1.2 can be expressed in a very convenient geometric form, noted in [8] and used very effectively in [21, 22]. We need a special class of positive measures on the plane which we describe next. Begin by choosing three unit vectors u, v, w in the plane such that $u+v+w = 0$. Consider the *lattice points* $iu + jv$ with integer coefficients i, j. A segment joining two nearest lattice points is called a *small edge*. We are interested in positive measures μ that are supported by a union of small edges, whose restriction to each small edge is a multiple of linear measure (arclength), and which satisfies the balance condition (called zero tension in [21])

$$\mu(AB) - \mu(AB') = \mu(AC) - \mu(AC') = \mu(AD) - \mu(AD') \qquad (1.1)$$

whenever A is a lattice point and the lattice points B, C', D, B', C, D' are in cyclic order around A. (Another way to define the objects of interest is to say they associate to each small edge e a nonnegative number $\mu(e)$ and that these numbers satisfy the balance condition (1.1) at every lattice point.) If e is a small edge, the value $\mu(e)$ is equal to the density of μ relative to linear measure on that edge.

Fix now an integer $r \geq 1$, and denote by $\triangle_r$ the (closed) triangle with vertices $0, ru$, and $ru + rv = -rw$. We will use the notation $A_j = ju, B_j = ru + jv$, and $C_j = (j - r)w$ for the lattice points on the boundary of $\triangle_r$. We also set

$$X_j = A_j + w, Y_j = B_j + u, Z_j = C_j + v, \quad j = 0, \ldots, r+1.$$

The following picture represents $\triangle_5$ and the points just defined; the labels are placed on the left.

Given a measure μ, a *branch point* is a lattice point incident to at least three edges in the support of μ. We only consider measures with at least one branch point. This excludes measures whose support consists of one or more parallel lines. We denote by $\mathcal{M}_r$ the collection of all measures μ satisfying the balance condition (1.1), whose branch points are contained in $\triangle_r$, and such that

$$\mu(A_j X_{j+1}) = \mu(B_j Y_{j+1}) = \mu(C_j Z_{j+1}) = 0, \quad j = 0, 1, \ldots, r.$$

The numbers

$$\mu(A_j X_j), \mu(B_j Y_j), \mu(C_j Z_j), \quad j = 0, 1, \ldots, r,$$

are called the *exit densities* of the measure $\mu \in \mathcal{M}_r$. The set $\mathcal{M}_r^{\mathbb{Z}} \subset \mathcal{M}_r$ consists of those measures μ which assign an integer density $\mu(e)$ to each small edge e. In particular, the exit densities of measures in $\mathcal{M}_r^{\mathbb{Z}}$ are nonnegative integers.

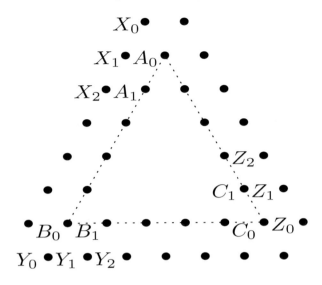

FIGURE 1.1:

The simplest nonzero measures in $\mathcal{M}_r$ are constructed as follows. Fix a lattice point $p \in \triangle_r$, and denote by ν_p the measure which assigns unit density to the half-lines $\{p + tu : t \geq 0\}$, $\{p + tv : t \geq 0\}$, and $\{p + tw : t \geq 0\}$, but assigns density zero to all small edges not contained in these half-lines. It is an easy exercise to show that the measures in $\mathcal{B}_r = \{\nu_p : p \in \triangle_r \text{ a lattice point}\}$ are linearly independent and every measure in $\mathcal{M}_r$ can be written as a linear combination of measures in $\mathcal{B}_r$ with real (not necessarily positive) coefficients. For instance, the measure whose support is pictured below is a combination of four measures from $\mathcal{B}_r$, one of them with a negative coefficient.

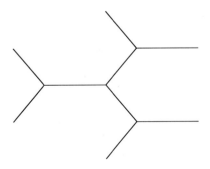

FIGURE 1.2:

Operator theory and Schubert calculus 13

This way of writing μ does not seem to be particularly useful, but it provides an immediate proof of the following result.

Proposition 1.3.1. *For any measure $\mu \in \mathcal{M}_r$ we have*

$$\sum_{j=0}^{r} \mu(A_j X_j) = \sum_{j=0}^{r} \mu(B_j Y_j) = \sum_{j=0}^{r} \mu(C_j Z_j).$$

Proof. The identity is trivially verified for $\mu = \nu_p$. The general case follows by linearity. $\square$

The common value of the sums in the preceding proposition is called the *weight* of the measure μ and it is denoted $\omega(\mu)$.

Suppose now that $\mu \in \mathcal{M}_r^{\mathbb{Z}}$. The weight $\omega(\mu)$ is an integer. We set $n = r + \omega(\mu)$ and define sets $I_\mu, J_\mu, K_\mu \subset \{1, \dots, n\}$ of cardinality r by $I_\mu = \{i_1, \dots, i_r\}$, where

$$i_x = x + \sum_{j=0}^{x-1} \mu(A_j X_j), \quad x = 1, \dots, r,$$

and J_μ, K_μ are defined by similar formulas, with $B_j Y_j$ and $C_j Z_j$, respectively, in place of $A_j X_j$.

Theorem 1.3.2. (The Littlewood–Richardson rule) *Given sets $I, J, K \subset \{1, \dots, n\}$ of cardinality r, the number c_{IJK} is the number of measures $\mu \in \mathcal{M}_r^{\mathbb{Z}}$ with weight $\omega(\mu) = n - r$ such that $I = I_\mu$, $J = J_\mu$, and $K = K_\mu$.*

The actual calculation of c_{IJK} can be quite lengthy. A. Buch wrote a software package for such calculations. The triples I, J, K for which $c_{IJK} = 1$ are of special interest and the measures which give rise to such triples are said to be *rigid* [22]. A nonzero rigid measure which is also minimal (relative to the ordering of measures) was called a *skeleton* in [5]. Every rigid measure in $\mathcal{M}_r^{\mathbb{Z}}$ is (uniquely) a linear combination of skeletons with positive integer coefficients. The measures ν_p are skeletons but there are far more complicated skeletons— a complete enumeration seems out of reach. We give one illustration of a fairly complex skeleton with weight 7. We tried to indicate edges with higher densities by thicker lines. The exit densities, enumerated counterclockwise and starting with the left side, are $1141 - 331 - 11311$.

The smallest example of sets I, J, K with $c_{IJK} > 1$ arises for $n = 6$, $r = 3$, and $I = J = K = \{2, 4, 6\}$. One can use Theorem 1.3.2 to calculate $c_{IJK} = 2$. It is an interesting linear algebra problem to solve the corresponding intersection problem. We are given spaces $E_2 \subset E_4 \subset \mathbb{C}^6$, $F_2 \subset F_4 \subset \mathbb{C}^6$, and $G_2 \subset G_4 \subset \mathbb{C}^6$ (subscripts indicate dimensions). We need to find a space $M_3 \subset \mathbb{C}^6$ which intersects E_2, F_2, G_2 in spaces of dimension at least 1, and the spaces E_4, F_4, G_4 in spaces of dimension at least 2. Generically, there should be two solutions corresponding to the two roots of a quadratic equation. (The

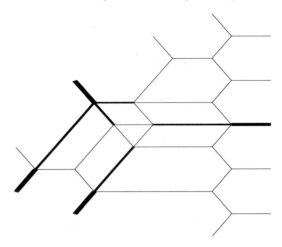

FIGURE 1.3:

equation in question is in fact the characteristic equation of a linear operator acting on a space of dimension 2. Every equation of this type can arise in this problem, showing in particular that there may be no solutions if real coefficients are used.)

Unlike the case $c_{IJK} > 1$, triples I, J, K with $c_{IJK} = 1$ yield intersection problems which can be solved over an arbitrary field of scalars. Given a field $\mathbb{F}$, the Grassmannian $G(n, r, \mathbb{F})$ and the Schubert varieties $\mathfrak{S}(\mathcal{E}, I)$ are defined as in the complex case, with $\mathbb{F}$ in place $\mathbb{C}$. The following result is proved in [5].

Theorem 1.3.3. *Let $\mathbb{F}$ be an arbitrary field, let $n > r > 0$ be integers, let $I, J, K \subset \{1, \ldots, n\}$ be sets of cardinality r such that $c_{IJK} = 1$, and let $\mathcal{E}, \mathcal{F}, \mathcal{G}$ be arbitrary complete flags in $\mathbb{F}^n$. Then the intersection*

$$\mathfrak{S}(\mathcal{E}, I) \cap \mathfrak{S}(\mathcal{F}, J) \cap \mathfrak{S}(\mathcal{G}, K)$$

is not empty.

The proof proceeds by producing an element in the intersection obtained (at least generically) by performing lattice operations, that is, sums and intersections, on the spaces in $\mathcal{E}$, $\mathcal{F}$, and $\mathcal{G}$. We illustrate this with two examples corresponding to the simplest types of skeletons. For the first example, take μ to be an integer multiple of ν_p for some lattice point $p \in \triangle_r$ and suppose that $\mu(A_i X_i)$, $\mu(B_j Y_j)$, and $\mu(C_k Z_k)$ are the three nonzero exit densities of ν. Then the solution of the intersection problem is generically $M = E_i + F_j + G_k$. In the nongeneric case, M has dimension less than r, and any space of dimension r containing M is a solution. For the second example, suppose that the support of μ is as pictured in Figure 1.2 and the nonzero exit densities of μ

Operator theory and Schubert calculus 15

are $\mu(A_{i_1}, X_{i_1})$, $\mu(A_{i_2} X_{i_2})$, $\mu(B_{j_1} Y_{j_1})$, $\mu(B_{j_2} Y_{j_2})$, $\mu(C_{k_1} Z_{k_1})$, and $\mu(C_{k_2} Z_{k_2})$, where $i_1 < i_2$, $j_1 < j_2$, and $k_1 < k_2$.

The solution of the corresponding intersection problem is given (generically) by the more complex formula

$$X + Y + Z,$$

where

$$
\begin{aligned}
X &= (E_{i_1} + F_{j_1} + G_{j_2}) \cap (E_{i_1} + F_{j_2} + G_{j_1}), \\
Y &= (E_{i_1} + F_{j_1} + G_{j_2}) \cap (E_{i_2} + F_{j_1} + G_{j_1}), \\
Z &= (E_{i_2} + F_{j_1} + G_{j_1}) \cap (E_{i_1} + F_{j_2} + G_{j_1}).
\end{aligned}
$$

Appropriate modifications need to be made in nongeneric cases: one may need to replace the sums by larger spaces with the appropriate dimensions and some of the intersections with smaller spaces with the appropriate dimensions. The complexity of these explicit formulas (as measured, for instance, by the number of nested parentheses) increases considerably for more complicated rigid measures μ.

1.4 Practical intersection theory

We explain how to find the unique element of $\mathfrak{S}(\mathcal{E}, I) \cap \mathfrak{S}(\mathcal{F}, J) \cap \mathfrak{S}(\mathcal{G}, K)$ when $c_{IJK} = 1$. As seen above, the sets I, J, K are obtained from a rigid measure and each rigid measure is a sum of integer multiples of skeletons. The simplest case would seem to be that of a multiple of a skeleton, but we have seen earlier that skeletons can be quite complicated. In order to deal with skeletons we use a duality observation: finding a space $M \subset \mathbb{C}^n$ of dimension r is equivalent to finding its orthogonal complement $M^\perp$ which has dimension $n - r$. (When working over an arbitrary field $\mathbb{F}$, the space $M^\perp$ should be the annihilator of M in the dual of $\mathbb{F}^n$.) Moreover, the statement that

$$M \in \mathfrak{S}(\mathcal{E}, I) \cap \mathfrak{S}(\mathcal{F}, J) \cap \mathfrak{S}(\mathcal{G}, K)$$

is equivalent to

$$M^\perp \in \mathfrak{S}(\widetilde{\mathcal{E}}, \widetilde{I}) \cap \mathfrak{S}(\widetilde{\mathcal{F}}, \widetilde{J}) \cap \mathfrak{S}(\widetilde{\mathcal{G}}, \widetilde{K}),$$

where sets of the form $\widetilde{I}$ have been defined earlier, and $\widetilde{\mathcal{E}}$ is the flag obtained by taking the orthogonal complements of the flags in $\mathcal{E}$. This seemingly trivial observation is useful because the dual intersection problem is given by a different measure. Indeed, if the sets I, J, K are produced by a measure $\mu \in \mathcal{M}_r^{\mathbb{Z}}$ it follows that $\widetilde{I}, \widetilde{J}, \widetilde{K}$ are produced by the *dual measure* $\mu^* \in \mathcal{M}_{n-r}^{\mathbb{Z}}$. Moreover, if μ is an integer multiple of a skeleton, then μ^* no longer has this property.

We illustrate this combinatorial fact (proved in [5]) for the skeleton pictured in Figure 1.2. Suppose that the nonzero exit densities of μ are $\mu(A_{i_1}, X_{i_1})$, $\mu(A_{i_2} X_{i_2})$, $\mu(B_{j_1} Y_{j_1})$, $\mu(B_{j_2} Y_{j_2})$, $\mu(C_{k_1} Z_{k_1})$, and $\mu(C_{k_2} Z_{k_2})$, where $i_1 < i_2$, $j_1 < j_2$, and $k_1 < k_2$, and the measure itself assigns density ρ to each edge in its support. Then the dual measure belongs to $\mathcal{M}_{2\rho}^{\mathbb{Z}}$ and its support is illustrated below.

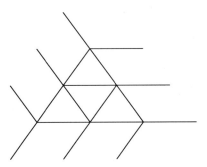

FIGURE 1.4:

This measure is seen to be a linear combination of six measures of the form ν_p (one for each branch point of the measure) with integer coefficients determined by the integers $i_1, i_2, j_1, j_2, k_1, k_2$, and r.

The considerations above show that the intersection problem associated with an integer multiple of a skeleton can be traded for a problem associated with a measure which is a sum of several distinct skeletons, each of them simpler (in a precise technical sense) than the original one. The solution of a problem associated with a linear combination of several skeletons can be obtained from the solutions of the problems associated with these skeletons. We illustrate the procedure with a sum of two measures of the form ν_p.

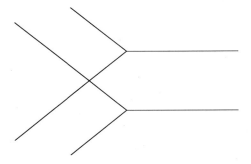

FIGURE 1.5:

Suppose again that the nonzero exit densities of μ are $\mu(A_{i_1}, X_{i_1})$, $\mu(A_{i_2} X_{i_2})$,

$\mu(B_{j_1}Y_{j_1})$, $\mu(B_{j_2}Y_{j_2})$, $\mu(C_{k_1}Z_{k_1})$, and $\mu(C_{k_2}Z_{k_2})$, where $i_1 < i_2$, $j_1 < j_2$, and $k_1 < k_2$. For simplicity, assign unit density to all edges in support of the measure, and note that $\mu = \mu_1 + \mu_2$, where μ_1 has nonzero exit densities $\mu(A_{i_1}, X_{i_1})$, $\mu(B_{j_1}Y_{j_1})$, and $\mu(C_{k_2}Z_{k_2})$, while μ_2 has nonzero exit densities $\mu(A_{i_2}X_{i_2})$, $\mu(B_{j_2}Y_{j_2})$, and $\mu(C_{k_1}Z_{k_1})$. The support of μ_1 intersects the support of μ_2 in exactly one point, and at that point the edges in support of μ_1 are $60°$ clockwise from those in support of μ_2. We say that μ_1 is *clockwise from* μ_2. In terms of intersection problems, clockwise measures must be treated first. What this means is that we have to follow the following three steps.

1. Solve the intersection problem associated with μ_1 (as if there were no μ_2). As seen earlier, the solution is simply $X = E_{i_1} + F_{j_1} + G_{k_2}$.

2. Intersect the flags $\mathcal{E}, \mathcal{F}, \mathcal{G}$ with X to form three new flags in X.

3. Solve the intersection problem associated with μ_2 *in the space* X. The solution is $(X \cap E_{i_2}) + (X \cap F_{j_2}) + (X \cap G_{k_1})$.

The space obtained in step (3) is (generically) the solution of the intersection problem associated with μ. Usually, a measure $\mu \in \mathcal{M}_r^{\mathbb{Z}}$ is a linear combination (with positive integer coefficients) of more than two skeletons $\mu_1, \ldots, \mu_q$. It is shown in [5] that we can assume that μ_{j+1} is *not* clockwise from μ_j for $j = 1, \ldots, q-1$. In this case the procedure above is followed inductively by solving the intersection problem for $\mu_1 + \cdots + \mu_j$, intersecting all the flags with the space thus obtained, and solving *in that space* the intersection problem for μ_{j+1}. The procedure may be tedious in practice, as the reader can verify by considering the measure illustrated in Figure 1.4. (One needs to put the six skeletons in some "clockwise" order. There are 90 ways to do this but any one of them will generically lead to the same final result.)

1.5 Back to operators

We return now to the three problems outlined in Section 1.1 and show how intersection theory applies to each of them.

1.5.1 Sums of Hermitian matrices

Suppose that $A, B \in M_n$ are Hermitian matrices. It follows easily from the considerations in Section 1.1.1 that the inequality

$$\sum_{k \in K} \lambda_k(A + B) \le \sum_{i \in I} \lambda_i(A) + \sum_{j \in J} \lambda_j(B) \tag{1.1}$$

18   *Recent Advances in Operator Theory and Operator Algebras*

holds whenever $c_{\tilde{I}\tilde{J}K} > 0$. Indeed, this condition ensures the existence of a projection P whose range satisfies the properties required for its proof. Klyachko [19] shows that the inequalities obtained this way are not only necessary but sufficient as well in the following precise sense.

Theorem 1.5.1. *Suppose the vectors $\alpha, \beta, \gamma \in \mathbb{R}^n$ are decreasing (that is, $\alpha_i \geq \alpha_{i+1}$, $\beta_i \geq \beta_{i+1}$, and $\gamma_i \geq \gamma_{i+1}$, $i = 1, \ldots, n$), satisfy the trace identity*

$$\sum_{j=1}^{n} (\alpha_j + \beta_j - \gamma_j) = 0,$$

and the inequalities

$$\sum_{k \in K} \gamma_k \leq \sum_{i \in I} \alpha_i + \sum_{j \in J} \beta_j$$

whenever $I, J, K \subset \{1, \ldots, n\}$ are sets of equal cardinality such that $c_{\tilde{I}\tilde{J}K} > 0$. Then there exist Hermitian matrices $A, B \in M_n$ such that $\Lambda(A) = \alpha$, $\Lambda(B) = \beta$, and $\Lambda(A + B) = \gamma$.

The arguments involve geometric invariant theory and we don't sketch them here. It was later shown by Belkale [1] that the system of inequalities in Theorem 1.5.1 is redundant. A minimal system of inequalities is obtained by restricting to those triples I, J, K with the property that $c_{\tilde{I}\tilde{J}K} = 1$. By Theorem 1.3.3, it is possible to provide a proof of the necessity of these inequalities which is entirely elementary. A version of Theorem 1.3.3 for II_1 factors, which is the real subject matter discussed in [5], allows one to obtain a complete characterization of the triples $(\lambda_A, \lambda_B, \lambda_{A+B})$ for selfadjoint elements in such a factor. These conditions consist of the trace identity

$$\int_0^1 (\lambda_A(t) + \lambda_B(t) - \lambda_{A+B}(t))\, dt = 0,$$

and a collection of inequalities of the form

$$\int_{\sigma_K} \lambda_{A+B}(t)\, dt \leq \int_{\sigma_I} \lambda_A(t)\, dt + \int_{\sigma_K} \lambda_B(t)\, dt,$$

where $I, J, K \subset \{1, \ldots, n\}$ for some integer n, I, J, K have cardinality $r < n$ with $c_{\tilde{I}\tilde{J}K} > 0$, and

$$\sigma_I = \bigcup_{i \in I} \left(\frac{i-1}{n}, \frac{i}{n} \right),$$

with similar definitions for σ_J and σ_K. The sufficiency of these conditions follows from Klyachko's theorem by a simple ultraproduct construction. The proof of necessity uses Belkale's reduction to the case $c_{\tilde{I}\tilde{J}K} = 1$ and the explicit construction of spaces in an intersection of Schubert cells in the Grassmann manifold associated with a factor $\mathcal{A}$. This is the set $G(n, r, \mathcal{A})$ consisting of those orthogonal projections $P \in \mathcal{A}$ with the property that $\tau(P) = r/n$.

Operator theory and Schubert calculus 19

Suppose now that A and B are positive compact operators on a Hilbert space $\mathcal{H}$. If P is a finite rank orthogonal projection on $\mathcal{H}$ we have $PAP + PBP = P(A + B)P$, so these operators of finite rank have to satisfy all the inequalities (1.1). By choosing sufficiently large projections P, we can ensure that the first N eigenalues of PAP, PBP, and $P(A + B)P$ coincide with the first N eigenvalues of A, B, and C, respectively. We conclude that $\Lambda(A), \Lambda(B)$, and $\Lambda(A + B)$ must also satisfy all the inequalities (1.1) when $c_{\widetilde{I}\widetilde{J}K} > 0$. It is not so obvious what to do about the trace identity unless A and B are trace class and

$$\sum_{j=1}^{\infty}(\lambda_j(A) + \lambda_j(B) - \lambda_j(A + B)) = 0.$$

It is however possible to replace this system of inequalities derived from Schubert calculus. The simplest of these inequalities is

$$\lambda_1(A) + \lambda_1(B) \leq \lambda_1(A + B) + \lambda_2(A + B). \tag{1.2}$$

Note that it involves one eigenvalue each for A and B, but *two* eigenvalues for $A + B$, so it is not entirely obvious that it is related to the inequalities derived from the intersection theory. It can however be easily obtained by taking traces in the identity $PAP + PBP = P(A + B)P$, where $P\mathcal{H}$ is generated by the eigenvectors for A and B corresponding to their highest eigenvalues. These "reverse Horn inequalities" become clearer when applied to compact operators that have negative as well as positive eigenvalues. Their necessity and sufficiency are discussed in [3], along with the still mysterious role played by the number 0 as an eigenvalue of A, B or $A + B$.

We would like to show how the inequality (1.2), and its analog for arbitrary compact operators arise from analogous inequalities in finite dimensions. Suppose that A is an arbitrary compact operator on a Hilbert space, and denote by $\Lambda(A)$ the double sequence

$$\lambda_1(A) \geq \lambda_2(A) \geq \cdots 0 \cdots \geq \lambda_{-2}(A) \geq \lambda_{-1}(A),$$

where $\{\lambda_n(A)\}_{n\in\mathbb{N}}$ lists the positive eigenvalues of A and $\{\lambda_{-n}(A)\}_{n\in\mathbb{N}}$ lists the negative eigenvalues of A, repeated according to multiplicity. As in the positive case, we set $\lambda_n(A) = 0$ for large n if A only has finitely many positive eigenvalues, with an analogous convention for the negative eigenvalues.

Suppose for the moment that $A, B \in M_n$. One inequality (which follows by subtracting one of the Horn inequalities from the trace identity) is as follows.

$$\lambda_1(A) + \lambda_n(A) + \lambda_1(B) + \lambda_n(B) \leq \lambda_1(A + B) + \lambda_2(A + B).$$

Note that the left-hand side contains the largest and the smallest eigenvalues of A and B. The corresponding inequality in the infinite-dimensional case is therefore

$$\lambda_1(A) + \lambda_{-1}(B) + \lambda_1(B) + \lambda_{-1}(B) \leq \lambda_1(A + B) + \lambda_2(A + B).$$

20   *Recent Advances in Operator Theory and Operator Algebras*

This is obtained, like the Horn inequalities, by compressing A and B to a sufficiently large space (dimension 6 will suffice) so that the relevant eigenvalues of the compressions of A, B and $A+B$ are precisely the ones appearing in this inequality. The inequality (1.2) follows in the case of nonnegative operators because $\lambda_{-1}(A) = \lambda_{-1}(B) = 0$ in that case. The other reverse Horn inequalities follow in the same manner, namely, by finding finite-dimensional Horn inequalities which depend in the right way on the dimension of the space.

We conclude this discussion of Hermitian sums by noting that the proof of the Horn conjecture in [21] also implies that, given $\alpha, \beta, \gamma \in \mathbb{R}^n$ satisfying the conditions of Theorem 1.5.1, the vectors α, β, γ satisfy a version of the Littlewood–Richardson rule involving measures in $\mathcal{M}_n$ rather than $\mathcal{M}_r^{\mathbb{Z}}$.

1.5.2   Products of matrices

Suppose that $T \in M_n$ and choose an orthonormal basis $\{e_1, \ldots, e_n\}$ in $\mathbb{C}^n$ with the property that $|T|e_j = \lambda_j(|T|)e_j$. Fix an integer $r \in \{1, \ldots, n\}$ and let $V, W : \mathbb{C}^r \to \mathbb{C}^n$ be two isometric operators. The connection between singular numbers and the intersection theory is made by determinants of the form $\det(W^*TV)$. More precisely, suppose that $I \subset \{1, \ldots, n\}$ contains n elements, and the flag $\mathcal{E}$ is determined by the condition that E_i is the span of $\{e_1, \ldots, e_i\}$. Choose the isometries V and W in such a way that $V\mathbb{C}^r \in \mathfrak{S}(\mathcal{E}, I)$ and $W\mathbb{C}^r \supset TV\mathbb{C}^r$. (The last inclusion must be an equality if T is invertible.) Then the inequality

$$|\det(W^*TV)| \geq \prod_{i \in I} \lambda_i(|T|)$$

holds. This follows from the observation that

$$\lambda_x(|W^*TV|) \geq \lambda_{i_x}(|T|), \quad x = 1, \ldots, r,$$

which is easily proved [4, Lemma 3.1].

The preceding observation is applied to products as follows. If $S, T \in M_n$ and $U, V, W : \mathbb{C}^r \to \mathbb{C}^n$ are isometric operators, we have

$$\det(W^*TSU^*) = \det((W^*TV)(V^*SU)) = \det(W^*TV)\det(V^*SU).$$

Suppose for the moment that T and S are invertible, $U\mathbb{C}^r = M$, $V\mathbb{C}^r = SM = N$, and $W\mathbb{C}^r = TSM = P$. Given three sets $I, J, K \subset \{1, \ldots, n\}$ of cardinality r such that $c_{IJK} > 0$ and flags $\mathcal{E}, \mathcal{F}, \mathcal{G}$, we can choose M in such a way that $M \in \mathfrak{S}(\mathcal{E}, I)$, $N \in \mathfrak{S}(\mathcal{F}, J)$, and $P \in \mathfrak{S}(\mathcal{G}, J)$. With appropriate choices of flags, the above determinant identity yields the Horn inequalities for the sequences $(\log |\lambda_i(T)|)_{i=1}^n$, $(\log |\lambda_i(S)|)_{i=1}^n$, and $(\log |\lambda_i(TS)|)_{i=1}^n$.

The case in which either T or S is not invertible can be treated as a limit of the invertible case. In the general case, it is best to write the Horn inequalities as inequalities between products of singular values so as to avoid infinite quantities. The sufficiency of the Horn inequalities (along with the

Operator theory and Schubert calculus 21

determinant identity

$$\prod_{i=1}^{n} \frac{|\Lambda_i(TS)|}{|\Lambda_i(T)||\Lambda_i(S)|} = 1)$$

is proved in [20] for the invertible case and in [4] for arbitrary T, S. Products of elements in a II_1 factor are also treated in [4].

1.5.3 Jordan models

The connection between the problem outlined in Section 1.1.3 and the intersection theory is easier to see in the more general context of finitely generated torsion modules over a discrete valuation ring. This simply is a commutative ring $\mathfrak{O}$ with no zero divisors in which there is an element p with the property that every proper ideal of $\mathfrak{O}$ is of the form $p^n \mathfrak{O}$ for some positive integer n. The finitely generated torsion modules over $\mathfrak{O}$ are of the form M/N, where M and N are finitely generated modules of the same rank. In other words, M and N are both isomorphic to $\mathfrak{O}^n$ for some positive integer n. Multiplication by p is a nilpotent endomorphism of such a module, and ordinary complex nilpotent matrices correspond to the special case in which $\mathfrak{O} = \mathbb{C}[[x]]$ and $p = x$. Finitely generated torsion modules are classified by an analog of the Jordan model obtained as follows. Let $L = M/N$ be such a module. Replacing L by an isomorphic module, we may assume that $M = \mathfrak{O}^n$ for some positive integer n, in which case $N = A\mathfrak{O}^n$, where A is an $n \times n$ matrix over $\mathfrak{O}$ satisfying $\det A \neq 0$. A further isomorphism allows us to assume that A is a diagonal matrix with diagonal entries $(p^{\alpha_1}, \ldots, p^{\alpha_n})$ for some integers $\alpha_1 \geq \cdots \geq \alpha_n \geq 0$. It follows immediately that L is isomorphic to a direct sum of cyclic modules:

$$\bigoplus_{i=1}^{n} \frac{\mathfrak{O}}{p^{\alpha_i}\mathfrak{O}}.$$

We denote by $\mathfrak{F}$ the field of quotients of $\mathfrak{O}$, that is, $\mathfrak{F} = \{a/b : a, b \in \mathfrak{O}, b \neq 0\}$. For any free module M over $\mathfrak{O}$ we denote by $\mathfrak{F}M = \mathfrak{F} \otimes_{\mathfrak{O}} M$ the corresponding vector space over $\mathfrak{F}$. In other words, if $\{e_1, \ldots, e_n\}$ is a basis of M over $\mathfrak{O}$ then it is a basis of $\mathfrak{F}M$ as a vector space over $\mathfrak{F}$. In general, it is convenient to identify M with $1M \subset \mathfrak{F}M$. With this identification, every $\mathfrak{F}$-linear subspace $V \subset \mathfrak{F}M$ determines a submodule $V \cap M$ of M. The rank of $V \cap M$ as a free module over $\mathfrak{O}$ is equal to $\dim_{\mathfrak{F}} V$.

We return now to our module $L = M/N = \mathfrak{O}^n/A\mathfrak{O}^n$, where A is a diagonal matrix with diagonal entries $(p^{\alpha_i})_{i=1}^{n}$ and $\alpha_1 \geq \cdots \geq \alpha_n \geq 0$. Denote by $\{e_1, \ldots, e_n\}$ the standard basis in M, and define a flag $\mathcal{E}$ in $\mathfrak{F}M$ by the requirement that E_i is the vector space generated by $\{e_1, \ldots, e_i\}, i = 1, \ldots, n$. Suppose that $I \subset \{1, \ldots, n\}$ is a set with $r < n$ elements, $V \in \mathfrak{S}(\mathcal{E}, I)$, and consider the free module $Q \subset M$ of rank r defined by $Q = V \cap M$. This module determines a submodule R of L by

$$R = \frac{Q+N}{N} \cong \frac{Q}{N \cap Q}.$$

The module L' is also a finitely generated torsion module over $\mathfrak{O}$ and is therefore isomorphic to

$$\bigoplus_{x=1}^{r} \frac{\mathfrak{O}}{p^{\beta_x}\mathfrak{O}}$$

for some integers $\beta_1 \geq \cdots \geq \beta_r \geq 0$. (There are only r integers in this sequence because Q has rank r.) The inequality

$$\sum_{x=1}^{r} \beta_x \geq \sum_{i \in I} \alpha_i$$

is proved in [7]. This inequality (and a related inequality for modules arising from a different Schubert variety) is the required link for application of the intersection theory. More precisely, given a submodule $L' \subset L$, we can write $L' = M'/N$ for some free module M' of rank n such that $N \subset M' \subset M$. Each of the three inclusions $N \subset M$, $N \subset M'$, and $M' \subset M$ is associated with a particular basis in which the inclusion operator is diagonal. One uses these bases to form three kinds of Schubert cells, and intersections of such Schubert cells yield certain modules Q of rank r. The Horn inequalities for the numbers α_i, and the corresponding numbers arising from L' and L/L', follow the consideration of these special modules Q. The spirit of the arguments is closely related to the multiplicative problem for singular values. The interested reader can find complete details in [7]. The sufficiency of the Horn inequalities in this problem follows from work of Klein [18].

The corresponding problem for nilpotent semigroups is also better explained in the more general context of operators of class C_0. The basic ideas are the same, but their execution becomes more technical. One of the reasons is that the ring H^∞ is far from being a discrete valuation ring and localization to a discrete valuation ring only works at points in the unit disk. The details cannot be summarized here, so we refer instead to [7].

References

[1] P. Belkale, Local systems on $\mathbb{P}^1 - S$ for S a finite set, *Compositio Math.* **129** (2001), 67–86.

[2] H. Bercovici, *Operator theory and arithmetic in H^∞*, Mathematical Surveys and Monographs, 26, American Mathematical Society, Providence, RI, 1988.

[3] H. Bercovici, H., W. S. Li, and D. Timotin, The Horn conjecture for sums of compact selfadjoint operators, *Amer. J. Math.* **131** (2009), 1543–1567.

[4] H. Bercovici, B. Collins, K. Dykema, and W. S. Li, Characterization of singular numbers of products of operators in matrix algebras and finite von Neumann algebras, *Bull. Sci. Math.* **139** (2015), no. 4, 400–419..

[5] H. Bercovici, B. Collins, K. Dykema, W. S. Li, and D. Timotin, Intersections of Schubert varieties and eigenvalue inequalities in an arbitrary finite factor, *J. Funct. Anal.* **258** (2010), 1579–1627.

[6] H. Bercovici, K. Dykema, and W. S. Li, The Horn inequalities for submodules, *Acta Sci. Math. (Szeged)* **79** (2013), 17–30.

[7] H. Bercovici and W. S. Li, Intersection theory and the Horn inequalities for invariant subspaces, *Acta Sci. Math. (Szeged)* **82** (2016), no. 1-2, 235–269.

[8] A. D. Berenshteĭn and A. V. Zelevinskiĭ, Involutions on Gel´fand-Tsetlin schemes and multiplicities in skew GL_n-modules (Russian), *Dokl. Akad. Nauk SSSR* **300** (1988), 1291–1294.

[9] A. S. Buch, The saturation conjecture (after A. Knutson and T. Tao). With an appendix by William Fulton, *Enseign. Math.* (2) **46** (2000), 43–60.

[10] C. Foias, On certain contraction semigroups, related with the representations of convolution algebras (Russian), *Rev. Roumaine Math. Pures Appl.* **7** (1962), 319–325.

[11] W. Fulton, *Intersection theory. Second edition*, Springer-Verlag, Berlin, 1998.

24 *References*

[12] ———, Eigenvalues, invariant factors, highest weights, and Schubert calculus, *Bull. Amer. Math. Soc. (N.S.)* **37** (2000), 209–249.

[13] J. Hersch and B. Zwahlen, Évaluations par défaut pour une somme quelconque de valeurs propres γ_k d'un opérateur $C = A + B$ à l'aide de valeurs propres α_i de A et β_j de B. (French), *C. R. Acad. Sci. Paris* **254** (1962), 1559–1561.

[14] A. Horn, Eigenvalues of sums of Hermitian matrices, *Pacific J. Math.* **12** (1962), 225–241.

[15] S. Johnson, *The Schubert calculus and eigenvalue inequalities for sums of Hermitian matrices*, Ph. D. thesis, University of California, Santa Barbara, 1979.

[16] S. L. Kleiman, Problem 15: rigorous foundation of Schubert's enumerative calculus, *Mathematical developments arising from Hilbert problems*, Amer. Math. Soc., Providence, RI, 1976, 445–482.

[17] T. Klein, The multiplication of Schur-functions and extensions of p-modules, *J. London Math. Soc.* **43** (1968), 280–284.

[18] ———, The Hall polynomial, *J. Algebra* **12** (1969), 61–78.

[19] A. Klyachko, Stable bundles, representation theory and Hermitian operators, *Selecta Math. (N.S.)* **4** (1998), 419–445.

[20] ———, Random walks on symmetric spaces and inequalities for matrix spectra, *Linear Algebra Appl.* **319** (2000), 37–59.

[21] A. Knutson and T. Tao, The honeycomb model of $GL_n(\mathbb{C})$ tensor products. I. Proof of the saturation conjecture, *J. Amer. Math. Soc.* **12** (1999), 1055–1090.

[22] A. Knutson, T. Tao, and C. Woodward, The honeycomb model of $GL_n(\mathbb{C})$ tensor products. II. Puzzles determine facets of the Littlewood–Richardson cone, *J. Amer. Math. Soc.* **17** (2004), 19-48.

[23] ———, A positive proof of the Littlewood–Richardson rule using the octahedron recurrence, *Electron. J. Combin.* **11** (2004), Research Paper 61, 18 pp.

[24] D. E. Littlewood, *The theory of group characters and matrix representations of groups*, Oxford University Press, New York, 1940.

[25] I. G. Macdonald, *Symmetric functions and Hall polynomials. Second edition. With contributions by A. Zelevinsky*, Oxford University Press, New York, 1995.

References

[26] H. Weyl, Das asymptotische Verteilungsgesetz der Eigenwerte linearer partieller Differentialgleichungen (mit einer Anwendung auf die Theorie der Hohlraumstrahlung) (German), *Math. Ann.* **71** (1912), 441-479.

Chapter 2

Non-selfadjoint operator algebras: dynamics, classification and C^*-envelopes

E.G. Katsoulis

2.1	Introduction	27
2.2	Examples	29
	2.2.1 The semicrossed product $C_0(X) \times_\sigma \mathbb{Z}^+$	29
	2.2.2 The classification problem for semicrossed products	32
	2.2.3 The tensor algebra of a graph $\mathcal{G}$	35
2.3	C^*-correspondences	37
	2.3.1 The gauge-invariance uniqueness theorems	44
2.4	Adding tails to a C^*-correspondence	45
	2.4.1 The Muhly–Tomforde tail	47
	2.4.2 The tail for (A, A, α)	47
2.5	The C^*-envelope of an operator algebra	51
	2.5.1 The C^*-envelope of an arbitrary operator algebra	56
2.6	Dynamics and classification of operator algebras	58
	2.6.1 Piecewise conjugate multisystems	60
	2.6.2 The multivariable classification problem	60
2.7	Crossed products of operator algebras	65
2.8	Local maps and representation theory	72

2.1 Introduction

This paper is an expanded version of the lectures I delivered at the Indian Statistical Institute, Bangalore, during the OTOA-2014 conference. My intention at the conference was to offer a gentle introduction to non-selfadjoint operator algebras, using topics that relate to my current research interests. One of my main goals was to provide to the audience most of the prerequisites for understanding the proof of Theorem 2.5.13, which identifies the C^*-envelope of a tensor algebra as the corresponding Cuntz–Pimsner C^*-algebra (providing these prerequisites meant of course that I had to survey a good deal of my operator algebra toolkit). Another goal was to demonstrate (through their classification) that certain non-selfadjoint algebras store a great deal of information about dynamical systems, in a much better way than their C*-counterparts do. These remain two of the main goals of these notes as well. During the preparation of this manuscript however, it occurred to me that there is something else that should be included here. Recently Chris Ramsey and I were able to extend the concept of a crossed product from C*-algebras to arbitrary operator algebras in such a way that many of the selfadjoint results are being preserved in this extra generality, e.g., Takai duality. This opens a new, exciting, and very promising area of research that somehow never attracted the attention it deserved, especially given the applications on C*-algebra theory. (See the discussion after Theorem 2.7.19 in Section 2.7.) Describing these developments has become yet another goal of these notes.

The paper is divided into eight sections, with this introduction being Section 2.1. In Section 2.2 several motivating examples of operator algebras are presented. In Section 2.3 we see the precise definitions for a C*-correspondence and the various operator algebras associated with it. We also state the various gauge invariance uniqueness theorems of Katsura and others. In Section 2.4 we describe a very general process for creating injective C*-correspondences from non-injective ones, without straying away from the associated Cuntz–Pimsner algebras (up to Morita equivalence). In Section 2.5 we give a complete treatment (including proofs) for the C*-envelope of a unital operator space. In Section 2.6 we describe various classification schemes for operator algebras coming from dynamical systems. In Section 2.7 we present a new topic that was not covered during the meeting at Bangalore and was mentioned earlier: crossed products of arbitrary operator algebras. The last section of the paper deals with local maps and gives us an opportunity to apply the concepts and tools developed so far to an area of study that goes back to the early work of Barry Johnson.

This paper is written in such a way that it could be used for self-study or as seminar notes for an introduction to non-selfadjoint operator algebras. Hence the first few sections unfold at a slower pace and they are less demanding compared to the last sections where I describe my current research. I have included as many proofs as possible in an article of this kind. Sometimes these are elementary and at other times only sketchy and covering a special case of the theorem. My intention is to give easy access to a variety of techniques without being fussy about completeness. Nevertheless I hope that the non-expert will look further into the details and the subject in general.

It goes without saying that this is not a comprehensive survey article and therefore many important topics and results are not being covered. The reader might also find that some of the topics covered receive an amount of attention which is perhaps unwarranted. This is only due to personal taste. I hope that these notes will serve as the nucleus for a forthcoming book on the topic that will do justice to both the subject and the mathematicians working on it.

Acknowledgment. I would like to express my gratitude to the organizers of OTOA-2014, and in particular to Jaydeb Sarkar, for the invitation to speak and also for the outstanding scientific environment, the hospitality, and the support they provided during my stay at the Indian Statistical Institute, Bangalore. It was a memorable experience that will stay with me for years to come.

2.2 Examples

In this section we examine various examples that motivate the general theory.

2.2.1 The semicrossed product $C_0(X) \times_\sigma \mathbb{Z}^+$

This is a very natural and important class of operator algebras. They were introduced by Arveson [3], and Arveson and Josephson [4], and their study was formalized by Peters [58] in 1984. Since then, these algebras have been investigated in one form or another by many authors [10, 11, 12, 13, 14, 19, 21, 39, 51, 64].

Let $X \subseteq \mathbb{C}$ be a locally compact Hausdorff space, $C_0(X)$ be the continuous functions on X, and $\sigma : X \to X$ a proper continuous map. One can give a "concrete" definition of $C_0(X) \times_\sigma \mathbb{Z}^+$ as follows.

Definition 2.2.1. *Let (X, σ) be as above. Given $x \in X$ and $f \in C_0(X)$, we define*

$$
\pi_x(f) = \begin{pmatrix} f(x) & 0 & 0 & \cdots \\ 0 & f(\sigma(x)) & 0 & \cdots \\ 0 & 0 & f(\sigma^{(2)}(x)) & \cdots \\ \vdots & \vdots & \vdots & \ddots \end{pmatrix}
$$

and

$$
S_x = \begin{pmatrix} 0 & 0 & 0 & \cdots \\ 1 & 0 & 0 & \cdots \\ 0 & 1 & 0 & \cdots \\ \vdots & \vdots & \vdots & \ddots \end{pmatrix}.
$$

In other words, for each $x \in X$, $f \in C_0(X)$

$$
\pi_x(f)\xi = (f(x)\xi_0, (f \circ \sigma)(x)\xi_1, (f \circ \sigma^{(2)})(x)\xi_2, \dots)
$$

and S_x is the forward shift

$$
S\xi = (0, \xi_0, \xi_1, \xi_2, \dots).
$$

The semicrossed product $C_0(X) \times_\sigma \mathbb{Z}^+$ is defined as the norm closed operator algebra acting on $\oplus_{x \in X} \mathcal{H}_x$ and generated by the operators

$$
\pi(f) \equiv \oplus_{x \in X} \pi_x(f) \text{ and } S\pi(f), f \in C_0(X),
$$

where

$$S \equiv \oplus_{x \in X} S_x.$$

Note the covariance relation

$$\pi(f)S = S\pi(f \circ \sigma), \quad f \in C_0(X).$$

A more "abstract" definition can be given as follows. Let $(\mathcal{A}, \alpha)$ be a C*-dynamical system, i.e., $\mathcal{A}$ is C*-algebra and $\alpha : \mathcal{A} \to \mathcal{A}$ is a non-degenerate *-endomorphism (an endomorphism that preserves approximate units).

Definition 2.2.2. *An isometric covariant representation (π, V) of the C*-dynamical system $(\mathcal{A}, \alpha)$ consists of a *-representation π of $\mathcal{A}$ on a Hilbert space $\mathcal{H}$ and an isometry $V \in B(\mathcal{H})$ so that*

$$\pi(A)V = V\pi(\alpha(A)), \quad \forall A \in \mathcal{A}.$$

Similar definitions apply for unitary or contractive covariant representations.

Definition 2.2.3. *Let $(\mathcal{A}, \alpha)$ be a C*-dynamical system. The algebra $\mathcal{A} \times_\alpha \mathbb{Z}^+$ is the universal operator algebra associated with "all" covariant representations of $(\mathcal{A}, \alpha)$, i.e., the universal algebra generated by a copy of $\mathcal{A}$ and an isometry V satisfying the covariant relations.*

Note that each covariant representation (π, V) provides a contractive representation $\pi \times V$ of $\mathcal{A} \times_\alpha \mathbb{Z}^+$. Similar definition can be given for the universal operator algebra $\mathcal{A} \times_\alpha^{un} \mathbb{Z}^+$ (resp. $\mathcal{A} \times_\alpha^{con} \mathbb{Z}^+$) associated with "all" unitary (resp. contractive) covariant representations.

The result below shows that the two definitions we have given so far for $C_0(X) \times_\sigma \mathbb{Z}^+$ actually describe the same object.

Theorem 2.2.4 (Peters [58]). *The representation $\oplus_{x \in X} \pi_x \times S_x$ of $C_0(X) \times_\sigma \mathbb{Z}^+$ is isometric.*

Proof. This result is an immediate consequence of the gauge-invariance uniqueness theorem of Katsura (Theorem 2.3.13). It pays however to revisit Peter's original argument, with σ being assumed a homeomorphism.

For each $x \in X$ and $f \in C_0(X)$ let

$$\rho_x(f)\xi = \left(\dots, (f \circ \sigma^{-1})(x)\xi_{-1}, f(x)\xi_0, (f \circ \sigma)(x)\xi_1, \dots \right)$$

and U_x is the forward shift on $l^2(\mathbb{Z})$. Let $\rho = \oplus_{x \in X} \rho_x$ and $U = \oplus_{x \in X} U_x$. It suffices to verify that $\rho \times U$ is isometric on $C_0(X) \times_\sigma \mathbb{Z}^+$ since it is the wot-limit of representations unitarily equivalent to $\oplus_{x \in X} \pi_x \times S_x$.

Because of the amenability of $\mathbb{Z}$, the representation $\rho \times U$ is faithful (and so isometric) for the C*-algebraic crossed product $C_0(X) \times_\sigma \mathbb{Z}$. Therefore for any *unitary* covariant representation (π, W) of (X, σ), the representation $\pi \times W$ is dominated in norm by $\rho \times U$. If (π, V') is an arbitrary covariant representation of (X, σ) and $V' = W + V$ its Wold decomposition, then the range space

$$I - P = W^*W = WW^*$$ commutes with π because of the covariance relations. So we may examine separately the covariant representations $((I-P)\pi, W)$ and $(P\pi, V)$. The previous considerations show that we only need to focus on the representation $(P\pi, V)$, with V a pure isometry, i.e., a direct sum of forward shifts. However in that case one can see that $P\pi \times V$ is unitarily equivalent to the restriction of a regular representation of the C*-algebra $C_0(X) \times_\sigma \mathbb{Z}$ and the conclusion follows. $\square$

It turns out that the representation theory of $C_0(X) \times_\sigma \mathbb{Z}^+$ allows more than just integrated forms of isometric covariant representations. Again this was first discovered by Peters in [58]. This result too has been extended by now in many different ways and it follows easily now from more general results. (See for instance [55].) However, the original argument and the result itself are still a source of inspiration for this author.

Theorem 2.2.5 (Peters [58]). *The algebras $\mathcal{A} \times_\alpha^{con} \mathbb{Z}^+$ and $\mathcal{A} \times_\alpha \mathbb{Z}^+$ (and $\mathcal{A} \times_\alpha^{un} \mathbb{Z}^+$ in the injective case) are isometrically isomorphic.*

Proof. I will prove only the isomorphism of $\mathcal{A} \times_\alpha^{un} \mathbb{Z}^+$, $\mathcal{A} \times_\alpha \mathbb{Z}^+$. Clearly the norm of a finite polynomial in $\mathcal{A} \times_\alpha \mathbb{Z}^+$ is dominated by its norm in $\mathcal{A} \times_\alpha^{con} \mathbb{Z}^+$, since any isometric covariant representation of $(\mathcal{A}, \alpha)$ is necessarily contractive.

To prove the converse, it suffices to show that any contractive covariant representation (π, T) of $(\mathcal{A}, \alpha)$ is the restriction on a co-invariant subspace of an isometric covariant representation $(\hat{\pi}, V_T)$ of $(\mathcal{A}, \alpha)$. The desired representation is actually given by the formula

$$\hat{\pi}(a) = \begin{pmatrix} \pi(a) & 0 & 0 & \cdots \\ 0 & \pi(\alpha(a)) & 0 & \cdots \\ 0 & 0 & \pi(\alpha^{(2)}(a)) & \cdots \\ \vdots & \vdots & \vdots & \ddots \end{pmatrix}, \quad a \in \mathcal{A}$$

and

$$V_T = \begin{pmatrix} T & 0 & 0 & \cdots \\ D_T & 0 & 0 & \cdots \\ 0 & I & 0 & \cdots \\ 0 & 0 & I & \cdots \\ \vdots & \vdots & \vdots & \ddots \end{pmatrix},$$

where $D_T = (I - T^*T)^{1/2}$. Verifying that the pair $(\hat{\pi}, V_T)$ satisfies the covariance relations means that we need to check equality at each entry to the left and right of the equation $\hat{\pi}(a)V_T = V_T\hat{\pi}(\alpha(a))$. This is immediate for all entries except from the $(2, 1)$ entry. There we need to show that

$$\pi(\alpha(a))D_T = D_T\pi(\alpha(a)), \ \forall a \in \mathcal{A}.$$

However,

$$\pi(\alpha(a))T^*T = T^*\pi(a)T = T^*T\pi(\alpha(a))$$

32 Recent Advances in Operator Theory and Operator Algebras

and this suffices. $\qquad\square$

Dilation results like the one above have received a lot of attention (and continue to do so) as they allow for a representation theory that is much richer than that of the selfadjoint theory. Nevertheless we will not be focusing on such results in these notes. See instead [10] for a recent comprehensive treatment.

2.2.2 The classification problem for semicrossed products

One of the fundamental problems in the study of algebras is their classification up to algebraic isomorphisms. The non-selfadjoint operator algebras are no exception to this rule

The problem of classifying the semicrossed products $C_0(X) \times_\sigma \mathbb{Z}^+$ as algebras was raised by Arveson [3]. Arveson's idea was to use non-selfadjoint crossed products in order to completely recover the dynamics. (This cannot be done using C*-algebraic crossed products.)

Assume that σ_1 and σ_2 are topologically conjugate, i.e., there exists a homeomorphism $\gamma : X_1 \to X_2$ so that

$$\gamma \circ \sigma_1 = \sigma_2 \circ \gamma.$$

Then the semicrossed products $C_0(X_1) \times_{\sigma_1} \mathbb{Z}^+$ and $C_0(X_2) \times_{\sigma_2} \mathbb{Z}^+$ are isomorphic as algebras. Hence conjugacy of the systems (X_1, σ_1) and (X_2, σ_2) is a sufficient condition for the isomorphism of $C_0(X_1) \times_{\sigma_1} \mathbb{Z}^+$ and $C_0(X_2) \times_{\sigma_2} \mathbb{Z}^+$. Is it necessary? Partial answers were given in the following cases:

- Both X_i are compact, both σ_i have no fixed points, plus some extra conditions. (Arveson and Josephson [4], 1969)

- Both X_i are compact and σ_i have no fixed points. (Peters [58], 1985)

- Both X_i are compact and the set

$$\{x \in X_i \mid \sigma_1(x) \neq x, \sigma_1^{(2)}(x) = \sigma_1(x)\}$$

  has empty interior. (Hadwin and Hoover [27], 1988)

- Both X_i are locally compact and σ_i are homeomorphisms. (Power [64], 1992)

The situation was finally resolved in 2008 by Davidson and the author. Our solution combined ideas of the previous authors regarding the character space with a new approach as to how to work with characters. Whereas the previous authors were using the character space in order to characterize various isomorphism-invariant ideals intrinsically, we instead used 2×2-upper triangular representations. The objective of using representations was to understand what happens to the character space under the action induced by an isomorphism at the algebra level. This approach also paid dividends in [11].

Let us look closely at the character space $\mathcal{M}_{(X,\sigma)}$ of the semicrossed product $C_0(X) \times_\sigma \mathbb{Z}^+$. Any $\theta \in \mathcal{M}_{(X,\sigma)}$ is determined uniquely by its values on $C_0(X)$ and $VC_0(X)$. Therefore there exist $x \in X$ and $\lambda \in \mathbb{C}$, $|\lambda| \leq 1$, so that $\theta \mid_{C_0(X)}$ is just evaluation on $x \in X$ and $\theta(Vg) = \lambda g(x)$, for any $g \in C_0(X)$. In that case we write $\theta = \theta_{x,\lambda}$ and we think of λ as the "value" of θ on V.[1] Note that in the case where x is not a fixed point for σ we have that $\lambda = 0$ and so there is only one choice for $\theta_{x,\lambda}$, i.e., $\theta_{x,0}$. Indeed this follows by applying θ to the covariance relation

$$ge_i V = e_i V(g \circ \sigma),$$

where $\{e_i\}_i$ is an approximate unit for $C_0(X)$ and $g \in C_0(X)$ satisfies $g(x) = 1$, $g(\sigma(x)) = 0$. If $x \in X$ is not a fixed point for σ, then there are other possibilities for $\theta_{x,\lambda}$ beyond $\theta_{x,0}$. The collection of all such characters $\theta_{x,\lambda}$ for a fixed $x \in X$ is denoted as $\mathcal{M}_{(X,\sigma,x)}$.

Now if $\mathcal{A}$ is an algebra, then $\mathrm{rep}_{\mathfrak{T}_2} \mathcal{A}$ will denote the collection of all representations of $\mathcal{A}$ onto $\mathfrak{T}_2$, the upper triangular 2×2 matrices. To each $\pi \in \mathrm{rep}_{\mathfrak{T}_2} \mathcal{A}$ we associate two characters $\theta_{\pi,1}$ and $\theta_{\pi,2}$ which correspond to compressions on the $(1,1)$ and $(2,2)$ entries, i.e.,

$$\theta_{\pi,i}(a) \equiv \langle \pi(a)\xi_i, \xi_i \rangle, \quad a \in \mathcal{A}, i = 1,2,$$

where $\{\xi_1, \xi_2\}$ is the canonical basis of $\mathbb{C}^2$. If $\gamma : \mathcal{A}_1 \to \mathcal{A}_2$ is an isomorphism between algebras, then γ induces isomorphisms,

$$\gamma_c : \mathcal{M}_{\mathcal{A}_1} \to \mathcal{M}_{\mathcal{A}_2} \qquad \text{by} \quad \gamma_c(\theta) = \theta \circ \gamma^{-1} \tag{2.1}$$

$$\gamma_r : \mathrm{rep}_{\mathfrak{T}_2} \mathcal{A}_1 \to \mathrm{rep}_{\mathfrak{T}_2} \mathcal{A}_2 \quad \text{by} \quad \gamma_r(\pi) = \pi \circ \gamma^{-1}, \tag{2.2}$$

which are compatible in the sense that

$$\gamma_c(\theta_{\pi,i}) = \theta_{\gamma_r(\pi),i}, \quad i = 1,2, \tag{2.3}$$

for any $\pi \in \mathrm{rep}_{\mathfrak{T}_2} \mathcal{A}_1$.

Let X be a locally compact Hausdorff space and σ a proper continuous map. For $x_1, x_2 \in X$, let

$$\mathrm{rep}_{x_1,x_2} C_0(X) \times_\sigma \mathbb{Z}^+ \equiv \{\pi \in \mathrm{rep}_{\mathfrak{T}_2} C_0(X) \times_\sigma \mathbb{Z}^+ \mid \theta_{\pi,i} \in \mathcal{M}_{(X,\sigma,x_i)}, i = 1,2\}.$$

Clearly, any element of $\mathrm{rep}_{\mathfrak{T}_2} C_0(X) \times_\sigma \mathbb{Z}^+$ belongs to $\mathrm{rep}_{x,y} C_0(X) \times_\sigma \mathbb{Z}^+$ for some $x, y \in X$.

Lemma 2.2.6. *Let X be a locally compact Hausdorff space and σ a proper continuous map on X. Assume that $x, y \in X$ are not fixed points for η and let $\pi \in \mathrm{rep}_{x,y} C_0(X) \times_\sigma \mathbb{Z}^+$. Then, $y = \sigma(x)$.*

---

[1] This statement is precise in the case where X is compact as V belongs to $C_0(X) \times_\sigma \mathbb{Z}^+$.

34   *Recent Advances in Operator Theory and Operator Algebras*

Proof. By assumption, $\theta_{\pi,1} = \theta_{x,0}$ and $\theta_{\pi,2} = \theta_{y,0}$ and so $\theta_{\pi,1}(gU) = \theta_{\pi,2}(gU) = 0$, for any $g \in C_0(\mathfrak{X})$. Therefore for each $g \in C_0(\mathfrak{X})$ there exists $c_g \in \mathbb{C}$ so that

$$\pi(gU) = \begin{pmatrix} 0 & c_g \\ 0 & 0 \end{pmatrix}.$$

Clearly there exists at least one $g \in C_0(\mathfrak{X})$ so that $c_g \neq 0$, or otherwise the range of π would be commutative. Applying π to $gUf = (f \circ \eta)gU$ for $f \in C_0(\mathfrak{X})$ and this particular g, we get

$$\begin{pmatrix} 0 & c_g \\ 0 & 0 \end{pmatrix} \begin{pmatrix} f(x) & t \\ 0 & f(y) \end{pmatrix} = \begin{pmatrix} f(\eta(x)) & t' \\ 0 & f(\eta(y)) \end{pmatrix} \begin{pmatrix} 0 & c_g \\ 0 & 0 \end{pmatrix}$$

for some $t, t' \in \mathbb{C}$, depending on f. By comparing $(1,2)$ entries, we obtain

$$f(y) = f(\eta(x)), \forall \in C_0(\mathfrak{X}),$$

i.e., $y = \eta(x)$, as desired. $\qquad\square$

Theorem 2.2.7 (Davidson and Katsoulis [14], 2008). *Let X_i be a locally compact Hausdorff space and let σ_i a proper continuous map on X_i. for $i = 1, 2$. Then the dynamical systems (X_1, σ_1) and (X_2, σ_2) are conjugate if and only if the semicrossed products $C_0(X_1) \times_{\sigma_1} \mathbb{Z}^+$ and $C_0(X_2) \times_{\sigma_2} \mathbb{Z}^+$ are isomorphic as algebras.*

Proof. I will prove the result under the assumption that both σ_i have no fixed points. In that case, $\mathcal{M}_{(X_i, \sigma_i, x)} = \{\theta_{x,0}\}$, $x \in X_i$, $i = 1, 2$, and so the character space $\mathcal{M}_{X_i, \sigma_i}$ equipped with the w^*-topology is homeomorphic to X_i, $i = 1, 2$.

Assume that there exists an isomorphism

$$\gamma : C_0(X_1) \times_{\sigma_1} \mathbb{Z}^+ \longrightarrow C_0(X_2) \times_{\sigma_2} \mathbb{Z}^+.$$

Then the map

$$\gamma_c : \mathcal{M}_{X_1, \sigma_1} \to \mathcal{M}_{X_2, \sigma_2}$$

provides a homeomorphism between the spaces X_1 and X_2. Furthermore

$$\gamma_r\big(\operatorname{rep}_{x, \sigma_1(x)} C_0(X) \times_\sigma \mathbb{Z}^+\big) = \operatorname{rep}_{\gamma_c(x), \gamma_c(\sigma_1(x))} C_0(X) \times_\sigma \mathbb{Z}^+.$$

The previous lemma applied to the right side of the above equation shows that

$$\sigma_2\big(\gamma_c(x)\big) = \gamma_c(\sigma_1(x))$$

and the conclusion follows. $\qquad\square$

2.2.3 The tensor algebra of a graph $\mathcal{G}$

This is a class of algebras that was introduced by Muhly and Solel [55] under the name quiver algebras. They generalize Popescu's non-commutative disc algebras [62, 63]. (See below for a precise definition.)

Let $\mathcal{G} = (\mathcal{G}^0, \mathcal{G}^1, r, s)$ be a countable directed graph and let $\mathcal{G}_\infty$ denote the (finite) path space of $\mathcal{G}$, i.e., all vertices $p \in \mathcal{G}^0$ and all finite paths

$$v = e_k e_{k-1} \ldots e_1$$

where the $e_i \in \mathcal{G}^1$ are edges satisfying $s(e_i) = r(e_{i-1})$, $i = 1, 2, \ldots, k$, $k \in \mathbb{N}$.

Let $\{\xi_v\}_{v \in \mathcal{G}_\infty}$ denote the usual orthonormal basis of the Fock space $\mathcal{H}_\mathcal{G} \equiv l^2(\mathcal{G}_\infty)$, where ξ_v is the characteristic function of $\{v\}$. The left creation operator L_v, $v \in \mathcal{G}_\infty$, is defined by

$$L_v \xi_w = \begin{cases} \xi_{qp} & \text{if } s(v) = r(w) \\ 0 & \text{otherwise.} \end{cases}$$

Definition 2.2.8. *The norm closed algebra generated by* $\{L_v \mid v \in \mathcal{G}_\infty\}$, *denoted as* $\mathcal{T}_\mathcal{G}^+$, *is the tensor algebra of the graph* $\mathcal{G}$. *Its weak closure, denoted as* $\mathcal{L}_\mathcal{G}$, *is the free semigroupoid algebra of* $\mathcal{G}$.

There is a particular class of graphs that deserves a special mention here. If $\mathcal{G}$ is the graph consisting of one vertex p and a loop e, then $L_p = I$ and the left creation operator L_e is unitarily equivalent to the forward shift on $l^2(\mathbb{N})$. For this graph $\mathcal{G}$, the tensor algebra $\mathcal{T}_\mathcal{G}^+$ is unitarily equivalent to the norm closed algebra generated by the shift operator, i.e., the disc algebra $A(\mathbb{D})$. More generally, if $\mathcal{G}$ is the graph consisting of one vertex p and $n \geq 2$ loop edges $e_1, e_2, \ldots, e_n$, then the corresponding creation operators $L_{e_1}, L_{e_2}, \ldots, L_{e_n}$ are pure isometries with orthogonal ranges whose joint wandering space is one-dimensional (Cuntz–Toeplitz isometries). In this case, the tensor algebra $\mathcal{T}_\mathcal{G}^+$ is called the non-commutative disc algebra and it is denoted as $\mathcal{A}_n$. These algebras were introduced by Popescu [62] and play an important role in the theory. As we shall see shortly, $\mathcal{A}_n$ is the universal algebra generated by a row isometry with n entries.

The algebras $\mathcal{T}_\mathcal{G}^+$ were classified by Kribs and the author in 2004.

Theorem 2.2.9 (Katsoulis and Kribs [45], 2004). *Let* $\mathcal{G}_1$, $\mathcal{G}_2$ *be directed graphs with no sinks. Then the tensor algebras* $\mathcal{T}_{\mathcal{G}_1}^+$ *and* $\mathcal{T}_{\mathcal{G}_2}^+$ *are isomorphic as algebras if and only if* $\mathcal{G}_1$ *are* $\mathcal{G}_2$ *are isomorphic as graphs.*

Once again a key ingredient of the proof is the use of multiplicative forms and 2×2-upper triangular representations. Indeed, in a graph algebra $\mathcal{T}_\mathcal{G}^+$ one identifies the vertices of the associated graph $\mathcal{G}$ with the connected components of the character space of $\mathcal{T}_\mathcal{G}^+$. The dimension of these connected components determines the number of loop edges supported on each one of the corresponding vertices. By looking at 2×2-upper triangular representations and the multiplicative forms appearing in the diagonal entries, one can decide

36 *Recent Advances in Operator Theory and Operator Algebras*

whether or not there exist arrows between the vertices corresponding to these forms. Calculating the number of these edges though requires an argument. See [45] for more details. Note the arguments in [45] actually inspired the work in [14].

Let's see how the above result can be used beyond operator algebra theory. We will reformulate a famous problem in graph theory as an operator algebra problem.

Definition 2.2.10. *Let $\mathcal{G}$ be a finite undirected graph with no loop edges or multiple edges between any two of its vertices. A vertex-deleted subgraph of $\mathcal{G}$ is a subgraph formed by deleting exactly one vertex from $\mathcal{G}$ and its incidence edges.*

Definition 2.2.11. *For a graph $\mathcal{G}$, the deck of $\mathcal{G}$, denoted as $D(\mathcal{G})$, is the multiset of all vertex-deleted subgraphs of $\mathcal{G}$. Each graph in $D(\mathcal{G})$ is called a card. Two graphs that have the same deck are said to be hypomorphic or reconstructions of each other.*

With these definitions at hand, we can state the following well-known conjecture.

Conjecture 2.2.12 (Kelly and Ulam). Any two hypomorphic graphs on at least three vertices have to be isomorphic.

A finite directed graph $\mathcal{G}$ will belong to the subclass $\mathfrak{G}_0$ of all directed graphs if $\mathcal{G}$ comes from a finite undirected graph by replacing each edge with two directed edges with opposite directions. The concepts of a card, a deck, and hypomorphism transfer to graphs in $\mathfrak{G}$ and the Reconstruction Conjecture can be stated as

Conjecture 2.2.13 (Reconstruction Conjecture of Kelly and Ulam). Any two hypomorphic graphs in $\mathfrak{G}_0$ on at least three vertices are necessarily isomorphic.

Definition 2.2.14. *If $\mathcal{G} \in \mathfrak{G}$, then a vertex-deleted subalgebra of $\mathcal{T}_{\mathcal{G}}^+$ is formed by deleting from $\mathcal{G}$ exactly one vertex and its incidence edges and then taking the subalgebra of $\mathcal{T}_{\mathcal{G}}^+$ formed by the partial isometries and projections corresponding to the remaining edges and vertices, respectively.*

Definition 2.2.15. *For a tensor algebra $\mathcal{T}_{\mathcal{G}}^+$, the deck of $\mathcal{T}_{\mathcal{G}}^+$, denoted as $D(\mathcal{T}_{\mathcal{G}}^+)$, is the multiset of all vertex-deleted subalgebras of $\mathcal{T}_{\mathcal{G}}^+$. Each graph in $D(\mathcal{T}_{\mathcal{G}}^+)$ is called a card. Two tensor algebras that have the same deck are said to be hypomorphic or reconstructions of each other.*

In an ongoing collaboration with Gunther Cornelissen at Utreht we have the following.

Theorem 2.2.16. *If $\mathcal{G}_1, \mathcal{G}_2 \in \mathfrak{G}_0$, then the graphs $\mathcal{G}_1$ and $\mathcal{G}_2$ are hypomorphic if and only if $\mathcal{T}_{\mathcal{G}_1}^+$ and $\mathcal{T}_{\mathcal{G}_2}^+$ are hypomorphic as operator algebras.*

Therefore the reconstruction conjecture admits the following equivalent form.

Non-selfadjoint operator algebras

Corollary 2.2.17. *The reconstruction conjecture in graph theory is equivalent to the assertion that hypomorphic tensor algebras of graphs in $\mathfrak{G}_0$ are necessarily isomorphic as algebras.*

There is a more abstract approach to defining the tensor algebra of a graph. The situation is similar to that of semicrossed products and there is a reason for that as we shall see soon.

Definition 2.2.18. *Let $\mathcal{G} = (\mathcal{G}^0, \mathcal{G}^1, r, s)$ be a countable directed graph. A family of partial isometries $\{L_e\}_{e \in \mathcal{G}^{(1)}}$ and projections $\{L_p\}_{p \in \mathcal{G}^{(0)}}$ is said to obey the Cuntz–Krieger–Toeplitz relations associated with $\mathcal{G}$ if and only if they satisfy*

$$(\dagger) \begin{cases} (1) & L_p L_q = 0 & \forall\, p, q \in \mathcal{G}^{(0)},\, p \neq q \\ (2) & L_e^* L_f = 0 & \forall\, e, f \in \mathcal{G}^{(1)},\, e \neq f \\ (3) & L_e^* L_e = L_{s(e)} & \forall\, e \in \mathcal{G}^{(1)} \\ (4) & L_e L_e^* \leq L_{r(e)} & \forall\, e \in \mathcal{G}^{(1)} \\ (5) & \sum_{r(e)=p} L_e L_e^* \leq L_p & \forall\, p \in \mathcal{G}^{(0)} \end{cases}$$

Definition 2.2.19. *The tensor algebra $\mathcal{T}_{\mathcal{G}}^+$ of a graph $\mathcal{G}$ is the universal operator algebra for all families of partial isometries $\{L_e\}_{e \in \mathcal{G}^{(1)}}$ and projections $\{L_p\}_{p \in \mathcal{G}^{(0)}}$ which obey the Cuntz–Krieger–Toeplitz relations associated with $\mathcal{G}$.*

It seems that we have given two different definitions for the same object. The next result shows that there are no contradictions arising from this.

Theorem 2.2.20 (Fowler, Muhly, and Raeburn [24], 2001)**.** *The representation of $\mathcal{T}_{\mathcal{G}}^+$ on the Fock space $\mathcal{H}_G$ is isometric.*

In particular, $\mathcal{A}_n$ is the universal algebra for a row isometry. The above result follows easily as a corollary of Theorem 2.3.13.

2.3   C*-correspondences

The algebras appearing in the previous sections are all examples of tensor algebras of C*-correspondences. In this section we describe in detail that broad class of operator algebras.

The construction of C*-algebras coming from a C*-correspondence originated in the seminal paper of Pimsner [60]. Pimsner considered only injective correspondences in [60]. Even though such correspondences are easier to work with, many natural C*-algebras come from non-injective ones. Many mathematicians tried to extend Pimsner's ideas to non-injective correspondences with varied levels of success. Nowadays Katsura's approach [48] is considered to be the most successful one; this is the one presented here. We start with the basic definitions.

Let A be a C*-algebra. An *inner-product right A-module* is a linear space X which is a right A-module together with a map

$$(\cdot,\cdot) \mapsto \langle\cdot,\cdot\rangle : X \times X \to A$$

such that

$$\langle \xi, \lambda\zeta + \eta \rangle = \lambda \langle \xi, \zeta \rangle + \langle \xi, \eta \rangle$$
$$\langle \xi, \eta a \rangle = \langle \xi, \eta \rangle\, a$$
$$\langle \eta, \xi \rangle = \langle \xi, \eta \rangle^*$$
$$\langle \xi, \xi \rangle \geq 0; \text{ if } \langle \xi, \xi \rangle = 0 \text{ then } \xi = 0.$$

For $\xi \in X$ we write $\|\xi\|_X^2 := \|\langle \xi, \xi \rangle\|_A$ and one can deduce that $\|\cdot\|_X$ is actually a norm. If X equipped with that norm is a complete normed space then it will be called *Hilbert A-module*. For a Hilbert A-module X we define the set $L(X)$ of the *adjointable maps* that consist of all maps $s : X \to X$ for which there is a map $s^* : X \to X$ such that

$$\langle s\xi, \eta \rangle = \langle \xi, s^*\eta \rangle, \ \xi, \eta \in X.$$

The compact operators $K(X) \subseteq L(X)$ is the closed subalgebra of $L(X)$ generated by the "rank one" operators

$$\theta_{\xi,\eta}(z) := \xi \langle \eta, z \rangle, \quad \xi, \eta, z \in X.$$

Definition 2.3.1. . *A C*-correspondence (X, A, φ) consists of a Hilbert A-module (X, A) and a non-degenerate left action*

$$\varphi : A \longrightarrow L(X).$$

If φ is injective then the C-correspondence (X, A, φ) is said to be injective.*

A (Toeplitz) representation (π, t) of X into a C*-algebra B is a pair of a *-homomorphism $\pi \colon A \to B$ and a linear map $t \colon X \to B$, such that

1. $\pi(a)t(\xi) = t(\varphi_X(a)(\xi))$,

2. $t(\xi)^*t(\eta) = \pi(\langle \xi, \eta \rangle_X)$,

for $a \in A$ and $\xi, \eta \in X$. An easy application of the C*-identity shows that

3. $t(\xi)\pi(a) = t(\xi a)$

is also valid. A representation (π, t) is said to be *injective* iff π is injective; in that case t is an isometry.

Definition 2.3.2. *The* Toeplitz–Cuntz–Pimsner *C*-algebra $\mathcal{T}_X$ of a C*-correspondence (X, A, φ) is the C*-algebra generated by all elements of the form $\pi_\infty(a), t_\infty(\xi)$, $a \in A$, $\xi \in X$, where (π_∞, t_∞) denotes the universal Toeplitz representation of (X, A, φ).*

Non-selfadjoint operator algebras 39

The ideas of Pimsner [60] were brought in the non-selfadjoint world by the pioneering work of Muhly and Solel [55], who recognized an important subalgebra of the Toeplitz–Cuntz–Pimsner C*-algebra $\mathcal{T}_X$.

Definition 2.3.3. *The* tensor algebra $\mathcal{T}_X^+$ *is the norm-closed subalgebra of* $\mathcal{T}_X$ *generated by all elements of the form* $\pi_\infty(a), t_\infty(\xi), a \in A, \xi \in X$, *where* (π_∞, t_∞) *denotes the universal Toeplitz representation of* (X, A, φ).

Let us see now how the examples of Section 2.2 manifest as tensor algebras of specific C*-correspondences.

Example 2.3.4. *(i) Let* $\mathcal{A}$ *be a unital C*-algebra and* α *a unital endomorphism. Set* $A = \mathcal{A}$, $X_\alpha = \mathcal{A}$,

$$\langle \xi, \eta \rangle = \xi^* \eta, \quad \xi, \eta \in X_\alpha$$

and $\varphi(a)\xi b = \alpha(a)\xi b$, *for* $a, b \in A$, $\xi \in X_\alpha$. *A non-degenerate[2] (Toeplitz) representation* (π, t) *of* (X_α, A) *into a C*-algebra* B *should satisfy*

$$1 = \pi(1) = \pi(\langle 1, 1 \rangle) = t(1)^* t(1),$$

where $1 \in \mathcal{A}$ *is the unit element. Furthermore Properties (i), (ii), and (iii) imply*

$$\begin{aligned} \pi(a)t(1) &= \pi(\varphi(a)1) = \pi(\alpha(a)) \\ &= \pi(1\alpha(a)) = t(1)\pi(\alpha(a)). \end{aligned}$$

This implies that the pair $(\pi, t(1))$ *is a covariant representation of the C*-dynamical system* $(\mathcal{A}, \alpha)$, *in the sense of Definition 2.2.2.*

Conversely if (π, V) *is a covariant representation of the C*-dynamical system* $(\mathcal{A}, \alpha)$, *then set* $t(\xi) = V\pi(\xi)$, $\xi \in \mathcal{A}$, *and verify that the pair* (π, t) *is a Toeplitz representation of* (X_α, A). *In this case the tensor algebra of* (X_α, A) *is the semicrossed product* $\mathcal{A} \times_\alpha \mathbb{Z}^+$.

(ii) Let $\mathcal{G} = (\mathcal{G}^0, \mathcal{G}^1, r, s)$ *be a finite graph. Let* $A = c_0(\mathcal{G}^0)$, $X_\mathcal{G} = c_0(\mathcal{G}^1)$

$$\langle \xi, \eta \rangle (p) = \sum_{s(e)=p} \overline{\xi(e)}\eta(e), \ \xi, \eta \in X_\mathcal{G}, \ p \in \mathcal{G}^0$$

and $(\varphi(f)\xi g)(e) = f(r(e))\xi(e)h(s(e))$, *with* $\xi \in X_\mathcal{G}$, $f, g \in A$ *and* $e \in \mathcal{G}^1$.

Let (π, t) *be a non-degenerate Toeplitz representation of the graph correspondence* $(X_\mathcal{G}, A)$. *Let* $L_p = \pi(1_p)$, $p \in \mathcal{G}^0$ *and* $L_e = t(1_e)$, $e \in \mathcal{G}^1$, *where* $1_e \in X_\mathcal{G}$ *denotes the characteristic function of the singleton* $\{e\}$, $e \in \mathcal{G}^1$, *and similarly for* $1_g \in A$, *with* $g \in \mathcal{G}^0$. *Then* $L_p L_q = \pi(1_p 1_q) = \delta_{p,q} L_p$, *where* $p, q \in \mathcal{G}^0$. *Also,*

$$L_e^* L_f = t(1_e)^* t(1_f) = \pi(\langle 1_e, 1_f \rangle) = 0, \ for \ e, f \in \mathcal{G}^1, e \neq f,$$

---

[2] If (π, t) is degenerate then restrict on the reducing subspace $\pi(1)$.

40   *Recent Advances in Operator Theory and Operator Algebras*

and similarly $L_e^* L_e = L_{s(e)}$, $e \in \mathcal{G}^1$. It is clear that the family $\{L_e\}_{e \in \mathcal{G}^{(1)}}$ of partial isometries and $\{L_p\}_{p \in \mathcal{G}^{(0)}}$ of projections obey the Cuntz–Krieger–Toeplitz relations of Definition 2.2.18.

Conversely, given a family $\{L_e\}_{e \in \mathcal{G}^{(1)}}$ of partial isometries and $\{L_p\}_{p \in \mathcal{G}^{(0)}}$ of projections obeying the Cuntz–Krieger–Toeplitz relations of Definition 2.2.18, we can define a Toeplitz representation of the graph correspondence $(X_{\mathcal{G}}, A)$ by setting $\pi(1_p) = L_p$, $p \in \mathcal{G}^0$, $t(1_e) = L_e$, $e \in \mathcal{G}^1$ and extending by linearity. In this case the tensor algebra of $(X_{\mathcal{G}}, A)$ is therefore $\mathcal{T}_{\mathcal{G}}^+$.

(iii) A broad class of C^*-correspondences arises naturally from the concept of a topological graph. A topological graph $\mathcal{G} = (\mathcal{G}^0, \mathcal{G}^1, r, s)$ consists of two σ-locally compact[3] spaces $\mathcal{G}^0$, $\mathcal{G}^1$, a continuous proper map $r : \mathcal{G}^1 \to \mathcal{G}^0$, and a local homeomorphism $s : \mathcal{G}^1 \to \mathcal{G}^0$. The set $\mathcal{G}^0$ is called the base (vertex) space and $\mathcal{G}^1$ the edge space. When $\mathcal{G}^0$ and $\mathcal{G}^1$ are both equipped with the discrete topology, we have a discrete countable graph (see below).

With a topological graph $\mathcal{G} = (\mathcal{G}^0, \mathcal{G}^1, r, s)$ there is a C^*-correspondence $X_{\mathcal{G}}$ over $C_0(\mathcal{G}^0)$. The right and left actions of $C_0(\mathcal{G}^0)$ on $C_c(\mathcal{G}^1)$ are given by

$$(fFg)(e) = f(r(e))F(e)g(s(e))$$

for $F \in C_c(\mathcal{G}^1)$, $f, g \in C_0(\mathcal{G}^0)$, and $e \in \mathcal{G}^1$. The inner product is defined for $F, G \in C_c(\mathcal{G}^1)$ by

$$\langle F \mid G \rangle (v) = \sum_{e \in s^{-1}(v)} \overline{F(e)} G(e)$$

for $v \in \mathcal{G}^0$. Finally, $X_{\mathcal{G}}$ denotes the completion of $C_c(\mathcal{G}^1)$ with respect to the norm

$$\|F\| = \sup_{v \in \mathcal{G}^0} \langle F \mid F \rangle (v)^{1/2}. \tag{2.1}$$

When $\mathcal{G}^0$ and $\mathcal{G}^1$ are both equipped with the discrete topology, then the tensor algebra $\mathcal{T}_{\mathcal{G}}^+ \equiv \mathcal{T}_{X_{\mathcal{G}}}^+$ associated with $\mathcal{G}$ coincides with the quiver algebra of Muhly and Solel [55]. In that case, $\mathcal{T}_{\mathcal{G}}^+$ has already been described.

Theorem 2.3.5. *Let (π, t) be a representation of a C^*-correspondence (X, A, φ). Then there exists a map*

$$\psi_t \colon K(X) \longrightarrow C^*(\pi, t)$$

so that $\psi_t(\theta_{\xi, \eta}) = t(\xi)t(\eta)^*$, for all $\xi, \eta \in X$. If π is injective then ψ_t is injective as well.

Proof. Represent $B \equiv C^*(\pi, t)$ faithfully on a Hilbert space $\mathcal{H}$, and consider the representation

$$\Phi \colon K(X) \longrightarrow B(X \otimes_A \mathcal{H}); K \longmapsto K \otimes \mathrm{id}.$$

---

[3]Due to this assumption, all discrete graphs appearing in this paper are countable.

Non-selfadjoint operator algebras 41

(See below for a precise definition of that tensor product.) It is easy to see now that we have an isometry $V(\xi \otimes h) = t(\xi)h \in \mathcal{H}$. Furthermore

$$V\Phi(\theta_{\xi,\eta}) = t(\xi)t(\eta)^*V. \tag{2.2}$$

However, $VV^*t(\eta)t(\xi)^* = t(\eta)t(\xi)^*$ and by taking adjoints we have $t(\xi)t(\eta)^*VV^* = t(\xi)t(\eta)^*$. Combining this with (2.2) we get

$$V\Phi(\theta_{\xi,\eta})V^* = t(\xi)t(\eta)^*VV^* = t(\xi)t(\eta)^*,$$

which clearly shows that $\psi_t(K) = V\Phi(K)V^*$, $K \in K(X)$, is the desired map. □

Definition 2.3.6. *A representation* (π, t) *of a* C*-correspondence (X, A, φ) *is said to be a covariant representation iff*

$$\pi(a) = \psi_t(\varphi(a)), \quad \text{for all } a \in J_X,$$

where $J_X = \varphi^{-1}(K(X)) \cap (\ker\varphi)^\perp$.

Definition 2.3.7. *The* Cuntz–Pimsner C*-algebra $\mathcal{O}_X$ *of a* C*-cor-break *respondence* (X, A, φ) *is the* C*-algebra generated by all elements of the form $\overline{\pi}_\infty(a), \overline{t}_\infty(\xi)$, $a \in A$, $\xi \in X$, where $(\overline{\pi}_\infty, \overline{t}_\infty)$ denotes the universal covariant representation of (X, A, φ).

It is not clear that a C*-correspondence admits non-trivial Toeplitz or covariant representations. For that purpose we introduce the interior tensor product of C*-correspondences.

The *interior* or *stabilized tensor product*, denoted by $X \otimes X$ or simply by $X^{\otimes 2}$, is the quotient of the vector space tensor product $X \otimes_{\text{alg}} X$ by the subspace generated by the elements of the form

$$\xi a \otimes \eta - \xi \otimes \varphi(a)\eta, \quad \xi, \eta \in X, a \in A.$$

It becomes a pre-Hilbert B-module when equipped with

$$(\xi \otimes \eta)a := \xi \otimes (\eta a),$$
$$\langle \xi_1 \otimes \eta_1, \xi_2 \otimes \eta_2 \rangle := \langle y_1, \varphi(\langle \xi_1, \xi_2 \rangle)\eta_2 \rangle$$

For $s \in L(X)$ we define $s \otimes \mathrm{id}_X \in L(X \otimes X)$ as the mapping

$$\xi \otimes y \mapsto s(\xi) \otimes y.$$

Hence $X \otimes X$ becomes a C*-correspondence by defining $\varphi_{X \otimes X}(a) := \varphi_X(a) \otimes \mathrm{id}_X$.

The *Fock space* $\mathcal{F}_X$ over the correspondence X is the interior direct sum of the $X^{\otimes n}$ with the structure of a direct sum of C*-correspondences over A,

$$\mathcal{F}_X = A \oplus X \oplus X^{\otimes 2} \oplus \dots.$$

42   *Recent Advances in Operator Theory and Operator Algebras*

Given $\xi \in X$, the (left) creation operator $t_\infty(\xi) \in \mathcal{L}(\mathcal{F}_X)$ is defined as

$$t_\infty(\xi)(a, \zeta_1, \zeta_2, \dots) = (0, \xi a, \xi \otimes \zeta_1, \xi \otimes \zeta_2, \dots).$$

For any $a \in A$, we define

$$\pi_\infty(a) = L_a \oplus \varphi(a) \oplus (\oplus_{n=1}^\infty \varphi(a) \otimes \mathrm{id}_n).$$

It is easy to verify that (π_∞, t_∞) is a Toeplitz representation of (X, A) which is called the *Fock representation* of (X, A). Note that π_∞ is faithful and so non-trivial.

We also need to produce non-trivial covariant representations. Our presentation follows that of Katsura [48]. We require the following.

Lemma 2.3.8. *Let (X, A) be a C*-correspondence and $J \subseteq A$ a closed ideal. If $k \in K(X)$, then the following are equivalent.*

(i) $k \in K(XJ) \equiv \left[\{ \theta_{\xi a, \eta} \mid \xi, \eta \in X, a \in J \} \right]$

(ii) $\langle k\xi, \eta \rangle \in J$, *for all* $\xi, \eta \in X$.

Proof. Since the operators satisfying (ii) form an ideal, it is enough to consider $k \geq 0$ satisfying (ii) and verify (i). It is enough to show that $k^3 \in K(XJ)$. Indeed, if $k = \lim_n \sum_i \theta_{\xi_i^n, \eta_i^n}$, then

$$k^3 = \lim_n \Big(\sum_i \theta_{\xi_i^n, \eta_i^n} \Big) k \Big(\sum_j \theta_{\xi_j^n, \eta_j^n} \Big)$$

$$= \lim_n \sum_{i,j} \theta_{\xi_i \langle \eta_i, K\xi_j \rangle, \eta_j} \in K(XJ)$$

as desired. $\qquad\qquad\square$

The previous lemma gives easily the following useful application.

Corollary 2.3.9. *Let (X, A), (Y, A) be C*-correspondences and let*

$$\varphi : A \longrightarrow L(Y)$$
$$\varphi_* : L(X) \longrightarrow L(X \otimes_\varphi Y); s \longmapsto s \otimes I.$$

Assume that $k \in K(X)$. Then $k \in \ker \varphi_*$ if and only if $k \in K(X \ker \varphi)$.

In particular, see Corollary below.

Corollary 2.3.10. *Let (X, A) be a C*-correspondence. Then for each $n \in \mathbb{N}$, the restricted map*

$$K(X^{\otimes n-1} J_X) \ni k \longmapsto k \otimes \mathrm{id} \in L(X^{\otimes n}) \tag{2.3}$$

is isometric.

Non-selfadjoint operator algebras   43

Proof. By the previous corollary, if k belongs to the kernel of (2.3), then

$$\langle k\overline{\xi}, \overline{\eta}\rangle \in \ker \varphi, \text{ for all } \overline{\xi}, \overline{\eta} \in X^{\otimes n-1}.$$

On the other hand $k \in K(X J_X)$ and so Lemma 2.3.8 shows that

$$\langle k\overline{\xi}, \overline{\eta}\rangle \in J_X \subseteq \ker \varphi^{\perp}, \text{ for all } \overline{\xi}, \overline{\eta} \in X^{\otimes n-1}.$$

Hence $k = 0$. $\qquad\qquad\qquad\qquad\qquad\qquad\qquad\qquad\qquad\qquad\qquad\square$

Let us prove now that (X, A) admits a non-trivial covariant representation. Note that for the Fock representation we have

$$\pi_{\infty}(a) - \psi_t\big(\varphi(a)\big) = L_a \oplus 0 \oplus 0 \oplus \dots, \quad a \in A.$$

Therefore if we want a covariant representation we need to somehow mod out with the ideal generated by the above differences, with a ranging over J_X. Let us try to determine that ideal.

Note that the above differences belong to $K(\mathcal{F}_X)$. Now if $\overline{\xi} \in X^{\otimes m}$ and $\overline{\eta} \in X^{\otimes n}$ and $a \in J_X$, then

$$t_{\infty}(\overline{\xi})(L_a \oplus 0 \oplus 0 \oplus \dots)t_{\infty}(\overline{\eta})^*$$
$$= \theta_{(0,0,\dots,\overline{\xi},\dots)a,(0,0,\dots,\overline{\eta}a,\dots)} \in K(\mathcal{F}_X J_X).$$

This calculation shows that the right candidate is $K(\mathcal{F}_X J_X)$. (Also note that the above implies that $K(\mathcal{F}_X J_X) \subseteq C^*(\pi_{\infty}, t_{\infty})$.) If we mod out by $K(\mathcal{F}_X J_X)$ we have a covariant representation $(\overline{\pi}_{\infty}, \overline{t}_{\infty})$ with

$$\overline{\pi}_{\infty} : A \xrightarrow{\pi_{\infty}} C^*(\pi_{\infty}, t_{\infty}) \xrightarrow{q} C^*(\pi_{\infty}, t_{\infty})/K(\mathcal{F}_X J_X)$$
$$\overline{t}_{\infty} : X \xrightarrow{\pi_{\infty}} C^*(\pi_{\infty}, t_{\infty}) \xrightarrow{q} C^*(\pi_{\infty}, t_{\infty})/K(\mathcal{F}_X J_X),$$

where q denotes the quotient map. In order to show that $(\overline{\pi}_{\infty}, \overline{t}_{\infty})$ is non-trivial, we verify that it is actually injective.

Assume that $\overline{\pi}_{\infty}(a) = 0$ and so $\pi_{\infty}(a) \in K(\mathcal{F}_X J_X)$. Then,

$$a \in J_X$$
$$\varphi(a) \in K(X J_X)$$
$$\varphi(a) \otimes \mathrm{id} \in K(X^{\otimes 2} J_X) \qquad\qquad\qquad (2.4)$$
$$\vdots$$

and also $\lim_n \|\varphi(a) \otimes \mathrm{id}^n\| = 0$, because $\pi_{\infty}(a)$ is compact. On the other hand, Corollary 2.3.10 implies that

$$\|\varphi(a)\| = \|\varphi(a) \otimes \mathrm{id}\| = \|\varphi(a) \otimes \mathrm{id}^2\| = \dots \qquad\qquad (2.5)$$

and so $\varphi(a) = 0$. Since $a \in J_X \subseteq \ker \varphi^{\perp}$ we obtain $a = 0$. This shows that $(\overline{\pi}_{\infty}, \overline{t}_{\infty})$ is injective and thus non-trivial.

44   *Recent Advances in Operator Theory and Operator Algebras*

2.3.1   The gauge-invariance uniqueness theorems

Having established that the representation theory of the Cuntz–Pimsner and Cuntz–Pimsner–Toeplitz algebras is not in vacuum, now we need a test to let us know when a particular representation of these algebras is actually faithful.

Definition 2.3.11. *A representation* (π, t) *of* X *is said to admit a gauge action if for each* $z \in \mathbb{T}$ *there exists a $*$-homomorphism*

$$\beta_z \colon \mathrm{C}^*(\pi, t) \to \mathrm{C}^*(\pi, t)$$

such that $\beta_z(\pi(a)) = \pi(a)$ *and* $\beta_z(t(\xi)) = zt(\xi)$, *for all* $a \in A$ *and* $\xi \in X$.

Theorem 2.3.12 (Katsura [49], 2004). *Let* (X, A, φ) *be a* C^**-correspondence and* (π, t) *a covariant representation that admits a gauge action and is faithful on* A. *Then the integrated representation* $\rho = \pi \times t$ *is faithful on* $\mathcal{O}_X$.

Proof. I will sketch the proof only in the case where $J_X = A$, i.e., φ is injective and $\varphi(A) \subseteq K(X)$.

Let $(\bar{\pi}_\infty, \bar{t}_\infty)$ be the universal covariant representation. Since both π and $\bar{\pi}_\infty$ are injective, the maps ψ_{t^n} and $\psi_{\bar{t}_\infty^n}$, $n \in \mathbb{N}$, are injective. Let

$$A_n \equiv \psi_{t^n}\big(K(X^{\otimes n})\big) \text{ and } \bar{A}_n \equiv \psi_{\bar{t}_\infty^n}\big(K(X^{\otimes n})\big), \quad n \in \mathbb{N}.$$

Since $A = J_X$ we have $\pi(A) \subseteq \psi_t(K(X))$ and inductively $A_n \subseteq A_{n+1}$, $n \in \mathbb{N}$. Similarly $\bar{A}_n \subseteq \bar{A}_{n+1}$, $n \in \mathbb{N}$. Set $A_\infty = \overline{\cup_{n=1}^\infty A_n}$ and similarly for $\bar{A}_\infty$. Clearly A_∞ and $\bar{A}_\infty$ are the fixed point algebras for the gauge actions on $\mathrm{C}^*(\pi, t)$ and $\mathcal{O}_X$, respectively. Since $\rho \mid_{\bar{A}_n} = \psi_{t^n} \circ \psi_{\bar{t}_\infty^n}^{-1}$, $n \in \mathbb{N}$, we obtain that $\rho \mid_{\bar{A}_n}$ is injective. Since $\bar{A}_\infty$ is an ascending union, $\rho \mid_{\bar{A}_\infty}$ is also injective. This suffices to prove the injectivity of ρ.

Indeed assume that there exists a positive $x \in \mathcal{O}_X \cap \ker \rho$. Since ρ intertwines the gauge actions, i.e., $\rho \circ \bar{\beta}_z = \beta_z \circ \rho$, $z \in \mathbb{T}$, we have

$$\rho \circ \bar{\Phi} = \Phi \circ \rho, \tag{2.1}$$

where Φ is the faithful expectation on $\mathrm{C}^*(\pi, t)$ projecting on A_∞, i.e., $\Phi(s) = \int \beta_z(s) dz$, $s \in \mathrm{C}^*(\pi, t)$, and similarly for $\bar{\Phi}$. Applying (2.1) on x we get that $\rho\big(\bar{\Phi}(x)\big) = 0$ and so $\bar{\Phi}(x) \in \ker \rho \cap \bar{A}_\infty = \{0\}$, as desired. $\qquad\square$

Combining the above theorem with our earlier results, we obtain that the representation $\bar{\pi}_\infty \times \bar{t}_\infty$ is faithful for $\mathcal{O}_X$, as $(\bar{\pi}_\infty, \bar{t}_\infty)$ is injective.

Theorem 2.3.13 (Katsura [49], 2004). *Let* (X, A, φ) *be a* C^**-correspondence and* (π, t) *a representation that admits a gauge action and satisfies*

$$I'(\pi, t) \equiv \{a \in A \mid \pi(a) \in \psi_t(K(X))\} = 0.$$

Then the integrated representation $\pi \times t$ *is faithful on* $\mathcal{T}_X$.

Non-selfadjoint operator algebras   45

Using the above theorem one can easily see that if (π_∞, t_∞) is the Fock representation, then $\pi_\infty \times t_\infty$ is faithful for $\mathcal{T}_X$.

Let us give an application of the gauge invariance uniqueness theorem to tensor algebras. We need the following result which generalizes the well-known fact that the restriction of the Calkin map on the algebra generated by the shift is an isometry. (Under the additional assumption that X is strict, this result was obtained by Muhly and Solel [55].)

Proposition 2.3.14 (Katsoulis and Kribs [46], 2006). *If (X, A, φ) is an injective correspondence, then*

$$\mathrm{alg}(\pi_\infty, t_\infty)/K(\mathcal{F}_X) \simeq \mathrm{alg}(\pi_\infty, t_\infty).$$

The following results tell us that for an injective correspondence (X, A), the tensor algebras $\mathcal{T}_X^+$ sit naturally inside $\mathcal{O}_X$. It generalizes the fact that the operator algebra generated by the shift operator is completely isometrically isomorphic with the disc algebra and therefore sits inside $C(\mathbb{T})$.

Corollary 2.3.15. *If (X, A) is an injective correspondence, then $\mathcal{T}_X^+$ embeds isometrically and canonically in $\mathcal{O}_X$.*

Proof. In the previous proposition we saw that $\mathrm{alg}(\pi_\infty, t_\infty)/K(\mathcal{F}_X) \simeq \mathrm{alg}(\pi_\infty, t_\infty)$ and so $\mathrm{alg}(\pi_\infty, t_\infty)/K(\mathcal{F}_{XJ_X}) \simeq \mathrm{alg}(\pi_\infty, t_\infty)$. However, as we commented right after the proof of Theorem 2.3.12,

$$\mathrm{alg}(\pi_\infty, t_\infty)/K(\mathcal{F}_{XJ_X}) \subseteq \mathrm{C}^*(\pi_\infty, t_\infty)/K(\mathcal{F}_{XJ_X}) \simeq \mathcal{O}_X$$

and we are done. $\qquad\qquad\qquad\qquad\qquad\qquad\qquad\qquad\qquad\qquad\qquad\square$

There is a stronger formulation for the above result: if (X, A) is an injective correspondence, then the C*-envelope of $\mathcal{T}_X^+$ is isomorphic to $\mathcal{O}_X$. The reader can go now to Section 2.5 for the appropriate definitions or even a proof of that result. Note however that the assumption of injectivity for (X, A) cannot be removed without the considerations of the section that follows.

2.4   Adding tails to a C*-correspondence

By adding a tail to a C*-correspondence, one can study arbitrary C*-correspondences with the aid of injective ones, which in general are better behaved. Loosely speaking, we say that a C*-correspondence (Y, B, ψ) arises from (X, A, φ) by adding a tail iff

(i) $X \subseteq Y$ and $A \subseteq B$, with

$$\psi(a)\xi = \varphi(a)\xi, \quad a \in A, \xi \in X$$

46   *Recent Advances in Operator Theory and Operator Algebras*

(ii) a covariant representation of (Y, B, ψ) restricts to a covariant representation of (X, A, φ)

(iii) $\mathcal{O}_X$ is a full corner of $\mathcal{O}_Y$.

The origins of this concept are in the theory of graph C*-algebras.

Let $\mathcal{G}$ be a connected, directed graph with a distinguished sink $p_0 \in \mathcal{G}^0$ and no sources. We assume that $\mathcal{G}$ is contractible at p_0, i.e., there exists a unique infinite path $w_0 = e_1 e_2 e_3 \ldots$ ending at p_0, i.e. $r(w_0) = p_0$ and the saturation of p is the whole graph (saturation in the sense of [5]). Let $p_n \equiv s(e_n)$, $n \geq 1$.

Let $(A_p)_{p \in \mathcal{G}^0}$ be a family of C*-algebras parameterized by the vertices of $\mathcal{G}$ so that $A_{p_0} = A$. For each $e \in \mathcal{G}^1$, we now consider a full, right Hilbert $A_{s(e)}$-module X_e and a $*$-homomorphism

$$\varphi_e \colon A_{r(e)} \longrightarrow \mathcal{L}(X_e)$$

satisfying the following requirements.

- If $e \neq e_1$, φ_e is injective and maps onto $\mathcal{K}(X_e)$.

- $\mathcal{K}(X_{e_1}) \subseteq \varphi_{e_1}(A)$ and

$$J_X \subseteq \ker \varphi_{e_1} \subseteq (\ker \varphi_X)^{\perp}. \tag{2.1}$$

- The maps φ_X and φ_{e_1} satisfy the *linking condition*

$$\varphi_{e_1}^{-1}(\mathcal{K}(X_{e_1})) \subseteq \varphi_X^{-1}(\mathcal{K}(X)). \tag{2.2}$$

Let

$$T_0 = c_0((A_p)_{p \in \mathcal{G}^0_-}),$$

where $\mathcal{G}^0_- \equiv \mathcal{G}^0 \setminus \{p_0\}$.

Let T_1 be the completion of $c_{00}((X_e)_{e \in \mathcal{G}^1})$ with respect to the inner product

$$\langle u, v \rangle (p) = \sum_{s(e)=p} \langle u_e, v_e \rangle, \quad p \in \mathcal{G}^0_-.$$

Equip T_1 now with a right T_0-action, so that

$$(ux)_e = u_e x_{s(e)}, \quad e \in \mathcal{G}^1, x \in T_0.$$

The pair (T_0, T_1) is the <u>tail</u> for (X, A, φ).

To the C*-correspondence (X, A, φ) and the data

$$\tau \equiv \left(\mathcal{G}, (X_e)_{e \in \mathcal{G}^1}, (A_p)_{p \in \mathcal{G}^0}, (\varphi_e)_{e \in \mathcal{G}^1} \right),$$

we now associate

$$\begin{aligned} A_\tau &\equiv A \oplus T_0 \\ X_\tau &\equiv X \oplus T_1 \end{aligned} \tag{2.3}$$

and we view X_τ as a A_τ-Hilbert module.

We define a left A_τ-action $\varphi_\tau : A_\tau \to \mathcal{L}(X_\tau)$ on X_τ by setting

$$\varphi_\tau(a, x)(\xi, u) = (\varphi_X(a)\xi, v),$$

where

$$v_e = \begin{cases} \varphi_{e_1}(a)(u_{e_1}), & \text{if } e = e_1 \\ \varphi_e(x_{r(e)})u_e, & \text{otherwise} \end{cases}$$

for $a \in A$, $\xi \in X$, $x \in T_0$, and $u \in T_1$.

Theorem 2.4.1 (Kakariadis and Katsoulis [40], 2012). *Let (X, A, φ) be a non-injective C*-correspondence and let X_τ be the graph C*-correspondence over A_τ defined above. Then X_τ is an injective C*-correspondence and the Cuntz–Pimsner algebra $\mathcal{O}_X$ is a full corner of $\mathcal{O}_{X_\tau}$.*

Furthermore, if (π, t) is a covariant representation of X_τ, then its restriction on X produces a covariant representation of (X, A, φ) .

2.4.1 The Muhly–Tomforde tail

Our Theorem 2.4.1 was inspired by relevant work of Muhly and Tomforde [56]. Given a (non-injective) correspondence (X, A, φ_X), Muhly and Tomforde construct the tail that results from the previous construction, with respect to data

$$\tau = \left(\mathcal{G}, (X_e)_{e \in \mathcal{G}^{(1)}}, (A_p)_{p \in \mathcal{G}^{(0)}}, (\varphi_e)_{e \in \mathcal{G}^{(1)}} \right)$$

defined as follows.

The graph $\mathcal{G}$ is illustrated in the figure below.

$A_p = X_e = \ker \varphi_X$, for all $p \in \mathcal{G}_-^{(0)}$ and $e \in \mathcal{G}^{(1)}$. Finally,

$$\varphi_e(a)u_e = au_e, \quad e \in \mathcal{G}^{(1)}, u_e \in X_e, a \in A_{r(e)}.$$

2.4.2 The tail for (A, A, α)

Given a (non-injective) correspondence (X, A, φ_X), we construct the tail that results from the previous construction, with respect to data

$$\tau = \left(\mathcal{G}, (X_e)_{e \in \mathcal{G}^{(1)}}, (A_p)_{p \in \mathcal{G}^{(0)}}, (\varphi_e)_{e \in \mathcal{G}^{(1)}} \right)$$

48   Recent Advances in Operator Theory and Operator Algebras

defined as follows.

Let $\theta : A \to M(\ker \varphi_X)$.

The graph $\mathcal{G}$ is once again

$$\bullet^{p_0} \xleftarrow{\quad e_1 \quad} \bullet^{p_1} \xleftarrow{\quad e_2 \quad} \bullet^{p_2} \xleftarrow{\quad e_3 \quad} \bullet^{p_3} \xleftarrow{\quad\quad} \bullet \xleftarrow{\quad\quad} \cdots$$

but $A_p = X_e = \theta(A)$, for all $p \in \mathcal{G}_-^{(0)}$ and $e \in \mathcal{G}^{(1)}$. Finally,

$$\varphi_e(a)u_e = \theta(a)u_e, \quad e \in \mathcal{G}^{(1)}, u_e \in X_e, a \in A_{r(e)}.$$

Using the technique of "adding tails" we can dispose of the injectivity assumption in Corollary 2.3.15.

Theorem 2.4.2 (Katsoulis and Kribs [46], 2006). *If (X, A, φ) is any C^*-correspondence, then $\mathcal{T}_X^+$ embeds isometrically and canonically in $\mathcal{O}_X$.*

Another application of adding tails appears in [40].

Theorem 2.4.3. *Let (A, α) be a C^*-dynamical system and X_α the pertinent correspondence. Then the Cuntz–Pimsner C^*-algebra $\mathcal{O}_{X_\alpha}$ is strongly Morita equivalent to a crossed product C^*-algebra.*

Finally let us give a sample of how exciting things can get with this process of "adding tails". This material is not required for accessing the rest of the paper.

Definition 2.4.4. *A multivariable C^*-dynamical system is a pair (A, α) consisting of a C^*-algebra A along with a tuple $\alpha = (\alpha_1, \alpha_2, \ldots, \alpha_n)$, $n \in \mathbb{N}$, of $*$-endomorphisms of A. The dynamical system is called injective iff $\cap_{i=1}^n \ker \alpha_i = \{0\}$. To the multivariable system (A, α) we associate a C^*-correspondence $(X_\alpha, A, \varphi_\alpha)$ as follows. Let $X_\alpha = A^n = \oplus_{i=1}^n A$ be the usual right A-module. That is,*

1. $(a_1, \ldots, a_n) \cdot a = (a_1 a, \ldots, a_n a)$,

2. $\langle (a_1, \ldots, a_n), (b_1, \ldots, b_n) \rangle = \sum_{i=1}^n \langle a_i, b_i \rangle = \sum_{i=1}^n a_i^* b_i$.

Also, by defining the $*$-homomorphism

$$\varphi_\alpha \colon A \longrightarrow \mathcal{L}(X_\alpha) \colon a \longmapsto \oplus_{i=1}^n \alpha_i(a),$$

X becomes a C^*-correspondence over A, with $\ker \varphi_\alpha = \cap_{i=1}^n \ker \alpha_i$ and $\varphi(A) \subseteq \mathcal{K}(X_\alpha)$.

It is easy to check that in the case where A and all α_i are unital, X is finitely generated as an A-module by the elements

$$e_1 := (1, 0, \ldots, 0), e_2 := (0, 1, \ldots, 0), \ldots, e_n := (0, 0, \ldots, 1),$$

where $1 \equiv 1_A$. In that case, (π, t) is a representation of this C*-correspondence if, and only if, the $t(\xi_i)$'s are isometries with pairwise orthogonal ranges and

$$\pi(c)t(\xi) = t(\xi)\pi(\alpha_i(c)), \quad i = 1, \ldots, n.$$

Definition 2.4.5. *The Cuntz–Pimsner algebra $\mathcal{O}_{(A,\alpha)}$ of a multivariable C*-dynamical system (A, α) is the Cuntz–Pimsner algebra of the C*-correspondence $(X_\alpha, A, \varphi_\alpha)$ constructed as above*

In the C*-algebra literature, the algebras $\mathcal{O}_{(A,\alpha)}$ are denoted as $A \times_\alpha \mathcal{O}_n$ and go by the name "twisted tensor products by $\mathcal{O}_n$". They were first introduced and studied by Cuntz [9] in 1981. In the non-selfadjoint literature, algebras are much more recent. In Section 2.6, we will see the tensor algebra $\mathcal{T}^+(A, \alpha)$ for a multivariable dynamical system (A, α). It turns out that $\mathcal{T}^+(A, \alpha)$ is completely isometrically isomorphic to the tensor algebra for the C*-correspondence $(X_\alpha, A, \varphi_\alpha)$. As such, $\mathcal{O}_{(A,\alpha)}$ is the C*-envelope of $\mathcal{T}^+(A, \alpha)$. Therefore, $\mathcal{O}_{(A,\alpha)}$ provides a very important invariant for the study of isomorphisms between the tensor algebras $\mathcal{T}_{(A,\alpha)}$.

We now apply our "adding tails" technique to the C*-correspondence defined above. The graph $\mathcal{G}$ that we associate with $(X_\alpha, A, \varphi_\alpha)$ has no loop edges and a single sink p_0. All vertices in $\mathcal{G}^{(0)} \backslash \{p_0\}$ emit n edges, i.e., as many as the maps involved in the multivariable system, and receive exactly one. In the case where $n = 2$, the following diagram illustrates $\mathcal{G}$.

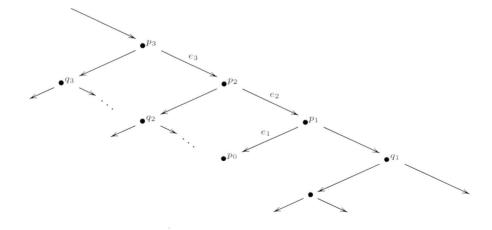

There is a unique infinite path w ending at p_0 whose saturation is the whole graph and so the requirements of Theorem 2.4.1 are satisfied, i.e., $\mathcal{G}$ is contractible at p_0.

Let $\mathcal{J} \equiv \cap_{i=1}^{n} \ker \alpha_i$ and let $M(\mathcal{J})$ be the multiplier algebra of $\mathcal{J}$. Let $\theta \colon A \longrightarrow M(\mathcal{J})$ the map that extends the natural inclusion $\mathcal{J} \subseteq M(\mathcal{J})$). Let $X_e = A_{s(e)} = \theta(A)$, for all $e \in \mathcal{G}^{(1)}$, and consider $(X_e, A_{s(e)})$ with the natural structure that makes it into a right Hilbert module.

For $e \in \mathcal{G}^{(1)} \backslash \{e_1\}$ we define $\varphi_e(a)$ as left multiplication by a. With that left action, clearly X_e becomes an $A_{r(e)} - A_{s(e)}$-equivalence bimodule. For $e = e_1$, it is easy to see that

$$\varphi_{e_1}(a)(\theta(b)) \equiv \theta(ab), \quad a, b \in A$$

defines a left action on $X_{e_1} = \theta(A)$, which satisfies both (2.1) and (2.2).

For the C*-correspondence $(X_\alpha, A, \varphi_\alpha)$ and the data

$$\tau = \left(\mathcal{G}, (X_e)_{e \in \mathcal{G}^{(1)}}, (A_p)_{p \in \mathcal{G}^{(0)}}, (\varphi_e)_{e \in \mathcal{G}^{(1)}} \right),$$

we now let $((X_\alpha)_\tau, A_\tau, (\varphi_\alpha)_\tau)$ be the C*-correspondence constructed as in the previous section. For notational simplicity $((X_\alpha)_\tau, A_\tau, (\varphi_\alpha)_\tau)$ will be denoted as $(X_\tau, A_\tau, \varphi_\tau)$. Therefore

$$A_\tau = A \oplus c_0(\mathcal{G}_-^{(0)}, \theta(A))$$
$$X_\tau = A^n \oplus c_0(\mathcal{G}^{(1)}, \theta(A)).$$

Now label the n-edges of $\mathcal{G}$ emitting from each $p \in \mathcal{G}_-^{(0)}$ as $p^{(1)}, p^{(2)}, \ldots, p^{(n)}$. It is easy to see now that the mapping

$$c_0(\mathcal{G}^{(1)}, \theta(A)) \ni u \longmapsto \oplus_{i=1}^{n} \{u(p^{(i)})\}_{p \in \mathcal{G}^{(0)}} \in \oplus_{i=1}^{n} c_0(\mathcal{G}_-^{(0)}, \theta(A))$$

establishes a unitary equivalence

$$X_\tau = A^n \oplus c_0(\mathcal{G}^{(1)}, \theta(A))$$
$$\cong \oplus_{i=1}^{n} \left(A \oplus c_0(\mathcal{G}_-^{(0)}, \theta(A)) \right)$$

between the Hilbert A-module X_τ and the n-fold direct sum of the C*-algebra $A \oplus c_0(\mathcal{G}_-^{(0)}, \theta(A))$, equipped with the usual $A \oplus c_0(\mathcal{G}_-^{(0)}, \theta(A))$-right action and inner product.

It only remains to show that the left action on X_τ comes from an n-tuple of *-endomorphisms of $A \oplus c_0(\mathcal{G}_-^{(0)}, \theta(A))$. This is established as follows.

For any $i = 1, 2, \ldots, n$ and $(a, x) \in A \oplus c_0(\mathcal{G}_-^{(0)}, \theta(A))$ we define

$$\hat{\alpha}_i(a, x) = (\alpha_i(a), \gamma_i(a, x))$$

where $\gamma_i(a, x) \in c_0(\mathcal{G}_-^{(0)}, \theta(A))$ with

$$\gamma_i(a, x)(p) = \begin{cases} \theta(a), & \text{if } p^{(i)} = e_0, \\ x(r(p^{(i)})), & \text{otherwise.} \end{cases}$$

It is easy to see now that $\left(A \oplus c_0(\mathcal{G}_-^{(0)}, \theta(A)), \hat{\alpha}_1, \ldots, \hat{\alpha}_n\right)$ is a multivariable dynamical system, so that the C*-correspondence associated with it is unitarily equivalent to $(X_\tau, A_\tau, \varphi_\tau)$.

We have therefore proved the result.

Theorem 2.4.6. *If (A, α) is a non-injective multivariable C*-dynamical system, then there exists an injective multivariable C*-dynamical system (B, β) so that the associated Cuntz–Pimsner algebras $\mathcal{O}_{(A,\alpha)}$ are full corner of $\mathcal{O}_{(B,\beta)}$. Moreover, if A belongs to a class $\mathcal{C}$ of C*-algebras which is invariant under quotients and c_0-sums, then $B \in \mathcal{C}$ as well. Furthermore, if (A, α) is non-degenerate, then so is (B, β).*

2.5   The C*-envelope of an operator algebra

Most experts will agree that the concept of the C*-envelope is at the heart of the modern operator algebra theory. This is one of the lasting contributions of Bill Arveson [1, 2] that will keep us busy for years to come. The presentation we give here is complete with proofs and it is essentially the Dritschel and McCullough approach to the subject [23] via maximal dilations.

What is the C*-envelope of an operator space $\mathcal{S}$? There is more than one way to approach the answer. Some might opt for the categorical approach: the C*-envelope is the "smallest" C*-algebra containing $\mathcal{S}$. (See the statement of Theorem 2.5.10.) Others, like myself, prefer the "utility grade" approach: the C*-envelope is the C*-algebra generated by the range of any maximal and isometric representation of $\mathcal{S}$. (Now *read* the proof of Theorem 2.5.10.) The truth be told, the C*-envelope is an elusive object as we only know of its existence through non-constructive proofs. As you can imagine, identifying the C*-envelope, even for very concrete algebras or spaces, can be quite a feat.

In this section, all operator spaces $\mathcal{S}$ satisfy $1 \in \mathcal{S} \subseteq C^*(\mathcal{S})$ and all completely contractive maps between operator spaces preserve the unit. See the monographs [7, 57] for the basic definitions and results, such as the one appearing below.

Theorem 2.5.1 (Arveson [2], 1969). *A (unital) completely contractive map $\varphi : \mathcal{S} \to B(\mathcal{H})$ admits a completely contractive (unital) extension*

$$\tilde{\varphi} : C^*(\mathcal{S}) \longrightarrow B(\mathcal{H}).$$

Definition 2.5.2. *A completely contractive (cc) map $\varphi : \mathcal{S} \to B(\mathcal{H})$ is said to have the unique extension property iff any completely contractive extension*

$$\tilde{\varphi} : C^*(\mathcal{S}) \longrightarrow B(\mathcal{H})$$

is multiplicative.

Definition 2.5.3. *If $\varphi_i : S \to B(\mathcal{H}_i)$, $i = 1, 2$, are cc maps then φ_2 is said to be a dilation of φ_1 (denoted as $\varphi_2 \geq \varphi_1$) if $\mathcal{H}_2 \supseteq \mathcal{H}_1$ and*

$$c_{\mathcal{H}_1}(\varphi_2(s)) \equiv P_{\mathcal{H}_1}\varphi_2(s)\mid_{\mathcal{H}_1} = \varphi_1(s), \quad \forall s \in S.$$

Definition 2.5.4. *A completely contractive (cc) map $\varphi : S \to B(\mathcal{H})$ is said to be maximal if it has no non-trivial dilations: $\varphi' \geq \varphi \implies \varphi' = \varphi \oplus \psi$ for some cc map ψ.*

Theorem 2.5.5 (Muhly and Solel [54], 1998). *A completely contractive map $\varphi : S \to B(\mathcal{H})$ is maximal iff it has the unique extension property.*

Proof. Suppose φ is maximal and $\tilde{\varphi}$ a cc extension on $C^*(S)$. By Stinespring's theorem there exists a dilation ρ of φ so that the diagram

$$
\begin{array}{ccc}
& & B(\mathcal{K}) \\
& \overset{\rho}{\nearrow} & \downarrow c \\
C^*(S) & \underset{\tilde{\varphi}}{\longrightarrow} & B(\mathcal{H})
\end{array}
$$

commutes. (Here c denotes compression on $\mathcal{H}$.) The map ρ is a *-representation and we may assume that

$$\big[\rho(C^*(S))(\mathcal{H})\big] = \mathcal{K}$$

or otherwise we compress.

Since $\tilde{\varphi}$ is maximal, for any $s \in S$ we have $\rho(s) = \varphi(s) \oplus s_1$ for some s_1 and so $\rho(s^*) = \rho(s)^* = \varphi(s)^* \oplus s_1^*$. Hence

$$\big[\rho(C^*(S))(\mathcal{H})\big] = \mathcal{H}$$

and so $\mathcal{K} = \mathcal{H}$ and $c = \text{id}$. Therefore $\tilde{\varphi} = \rho$ is multiplicative.

Conversely, assume that $\varphi : S \to B(\mathcal{H})$ has the unique extension property and let $\rho : S \to B(\mathcal{K})$ be a dilation of φ. Extend both φ and ρ to completely contractive maps $\tilde{\varphi}$ and $\tilde{\rho}$ so that the diagram

$$
\begin{array}{ccc}
& & B(\mathcal{K}) \\
& \overset{\tilde{\rho}}{\nearrow} & \downarrow c \\
C^*(S) & \underset{\tilde{\varphi}}{\longrightarrow} & B(\mathcal{H})
\end{array}
$$

commutes on S. Hence the completely contractive map $c \circ \tilde{\rho}$ agrees with φ on S and since φ has the unique extension property, $c \circ \tilde{\rho}$ is multiplicative. Hence

$$P_{\mathcal{H}}\tilde{\rho}(s^*s)P_{\mathcal{H}} = P_{\mathcal{H}}\tilde{\rho}(s^*)P_{\mathcal{H}}\tilde{\rho}(s)P_{\mathcal{H}}.$$

Also by the Swarchz inequality

$$P_{\mathcal{H}}\tilde{\rho}(s^*s)P_{\mathcal{H}} \geq P_{\mathcal{H}}\tilde{\rho}(s^*)\tilde{\rho}(s)P_{\mathcal{H}}.$$

Subtracting the above gives

$$0 \geq P_{\mathcal{H}}\tilde{\rho}(s^*)(I - P_{\mathcal{H}})\tilde{\rho}(s)P_{\mathcal{H}} = \left((I - P_{\mathcal{H}})\tilde{\rho}(s)P_{\mathcal{H}}\right)^*\left((I - P_{\mathcal{H}})\tilde{\rho}(s)P_{\mathcal{H}}\right) \geq 0$$

and so $\tilde{\rho}((\mathcal{S})$ leaves $\mathcal{H}$ invariant. Analogous calculations with $\tilde{\rho}(ss^*)$ also imply invariance of $\mathcal{H}$ by $\tilde{\rho}(\mathcal{S}^*) = \tilde{\rho}(\mathcal{S})^*$ and so $\mathcal{H}$ reduces $\tilde{\rho}(\mathcal{S})$ and therefore $\rho(\mathcal{S})$. Hence ρ is a trivial dilation. $\square$

As I explained in the introduction, the range of a completely isometric and maximal map will give us the C*-envelope. Proving that such maximal maps do exist requires a clever trick.

Theorem 2.5.6 (Dritschel and McCullough [23], 2005). *Every cc map $\varphi :$ $\mathcal{S} \to B(\mathcal{H})$ can be dilated to a maximal cc map $\varphi' : \mathcal{S} \to B(\mathcal{H}')$.*

Proof. For convenience we assume that both $\mathcal{A}$ and $\mathcal{H}$ are separable. The proof proceeds in two steps. First we prove that φ admits a dilation ψ on a Hilbert space $\mathcal{H}_\psi$ which is *maximal with respect to* φ. That means that any dilation ψ' of ψ on some Hilbert space $\mathcal{H}_{\psi'}$ satisfies

$$\left(P_{\mathcal{H}_{\psi'}} - P_{\mathcal{H}_\psi}\right)\psi' P_{\mathcal{H}} = P_{\mathcal{H}}\psi\left(P_{\mathcal{H}_{\psi'}} - P_{\mathcal{H}_\psi}\right) = 0. \tag{2.1}$$

By way of contradiction assume that such a dilation does not exist for φ. Hence there exists a dilation ψ_1 for φ which is non-trivial, i.e., it does not reduce $\mathcal{H}$. Since ψ_1 is not maximal with respect to φ, it admits a dilation ψ_2 that fails (2.1), i.e.,

$$\left(P_{\mathcal{H}_{\psi_2}} - P_{\mathcal{H}_{\psi_1}}\right)\psi' P_{\mathcal{H}} = P_{\mathcal{H}}\psi\left(P_{\mathcal{H}_{\psi_2}} - P_{\mathcal{H}_{\psi_1}}\right) = 0.$$

Since ψ_2 is not maximal with respect to φ, it admits a dilation ψ_3 that once again fails (2.1), i.e.,

$$\left(P_{\mathcal{H}_{\psi_3}} - P_{\mathcal{H}_{\psi_2}}\right)\psi' P_{\mathcal{H}} = P_{\mathcal{H}}\psi\left(P_{\mathcal{H}_{\psi_3}} - P_{\mathcal{H}_{\psi_2}}\right) = 0.$$

Continuing like this and using transfinite induction, we obtain dilations $\{\psi_i\}_{i\in\mathbb{I}}$ that fail (2.1) with respect to their successors, where $\mathbb{I}$ denotes the first uncountable ordinal. Since the gap ordinals are uncountable, we obtain a contradiction because of our separability assumption.

In order to finish the proof, let ψ_1 be a dilation of φ which is maximal with respect to φ, let ψ_2 be a dilation of ψ_1 which is maximal with respect to ψ_1, and so on. Any weak limit of the sequence $\{\psi_n\}_{n\in\mathbb{N}}$ gives the desired maximal dilation φ' of φ. $\square$

Notice an important implication of Theorem 2.5.5. If $\varphi : \mathcal{S} \to B(\mathcal{H})$ is completely contractive homomorphism of a (unital) operator algebra $\mathcal{S}$, then any maximal dilation of φ is automatically multiplicative. The same conclusion on multiplicativity holds for arbitrary completely isometric maps between operator algebras as we are about to see. First we need the following.

Lemma 2.5.7 (Arveson [2], 1969). *Let* $\mathcal{S}, \mathcal{T}$ *be operator spaces and* $\alpha : \mathcal{S} \longrightarrow \mathcal{T}$ *be a completely isometric (unital) surjection. If* $\varphi : \mathcal{T} \to B(\mathcal{H})$ *is maximal then*

$$\varphi \circ \alpha : \mathcal{S} \to B(\mathcal{H})$$

is also maximal.

Proof. Indeed assume that ρ dilates $\varphi \circ \alpha$ so that the diagram

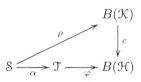

commutes. Then for any $t \in \mathcal{T}$ we have

$$P_{\mathcal{H}}\rho(\alpha^{-1}(t))\mid_{\mathcal{H}} = c(\rho(\alpha^{-1}(t))) = \varphi(t)$$

and so by the maximality of φ, there exists t_1 so that

$$\rho(\alpha^{-1}(t)) = \varphi(t) \oplus t_1.$$

Substituting $t = \alpha(s)$, $s \in \mathcal{S}$ in the above, we get

$$\rho(s) = (\varphi \circ \alpha)(s) \oplus s_1$$

and so $\varphi \circ \alpha$ is a maximal map. $\square$

Corollary 2.5.8. *If* $\mathcal{A}, \mathcal{B}$ *are unital operator algebras and*

$$\alpha : \mathcal{A} \longrightarrow \mathcal{B}$$

is a complete surjective isometry, then α *is multiplicative.*

Proof. Consider the diagram

$$\mathcal{A} \xrightarrow{\alpha} \mathcal{B} \xrightarrow{\rho} B(\mathcal{H})$$

with ρ a maximal completely isometric map. Since ρ is the restriction of a $*$-homomorphism on $C^*(\alpha(\mathcal{A}))$, ρ is multiplicative. By the previous lemma, $\rho \circ \alpha$ is maximal. Arguing as above, we obtain that $\rho \circ \alpha$ is also multiplicative. Hence, $\alpha = \rho^{-1} \circ (\rho \circ \alpha)$ is multiplicative. $\square$

The corollary above has an interesting consequence. If somehow the operator space $\mathcal{S}$ we are working with happens to be an operator algebra, then all completely isometric maps turn out to respect multiplication provided that the range is an algebra. In other words we can switch categories painlessly. Of course the standing assumption of unitality remains as Corollary 2.5.8 fails in the non-unital case.

The following results were discovered by Arveson [1] in special cases and by Hamana [26] in complete generality.

Theorem 2.5.9 (Hamana [26], 1979). *Let $\varphi : \mathcal{S} \to B(\mathcal{H})$ be a completely isometric maximal map. If $\mathcal{J} \subseteq C^*(\mathcal{S})$ is an ideal so that the quotient map*

$$q : C^*(\mathcal{S}) \longrightarrow C^*(\mathcal{S})/\mathcal{J}$$

is completely isometric on $\mathcal{S}$, then

$$\mathcal{J} \subseteq \ker \tilde{\varphi}$$

where $\tilde{\varphi}$ is the unique cc extension of φ to $C^*(\mathcal{S})$. (The ideal $\ker \tilde{\varphi}$ is said to be the Shilov ideal of $\mathcal{S} \subseteq C^*(\mathcal{S})$.)

Proof. Consider a map θ that makes the following diagram

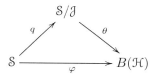

commutative. Now extend θ to obtain a diagram

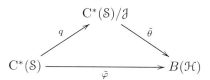

which is not a priori commutative. However, $\tilde{\theta} \circ q$ is a completely contractive extension of φ, which has the unique extension property. Hence $\tilde{\varphi} = \tilde{\theta} \circ q$. In particular, $\mathcal{J} = \ker q \subseteq \ker \tilde{\varphi}$. □

Finally here is the existence of the C^*-envelope.

Theorem 2.5.10 (Hamana [26], 1979). *Let $\mathcal{S}$ be a unital operator space. Then there exists a C^*-algebra $C^*_{env}(\mathcal{S})$ and a complete unital isometry*

$$j : \mathcal{S} \longrightarrow C^*_{env}(\mathcal{S})$$

so that for any other completely isometric unital embedding

$$\varphi : \mathcal{S} \longrightarrow \mathcal{C} = C^*(\varphi(\mathcal{S}))$$

we have $*$-homomorphism $\pi : \mathcal{C} \to C^*_{env}(\mathcal{S})$ so that $\pi \circ \varphi = j$.

Proof. The candidate for $C^*_{env}(\mathcal{S})$ is the pair $(C^*(j(\mathcal{S})), j)$ where j is any maximal and completely isometric representation j of $\mathcal{S}$, e.g., any maximal dilation of a completely isometric representation of $\mathcal{S}$.

Indeed if

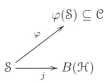

are as in the statement, then $j \circ \varphi^{-1}$ is a maximal map and so extends to a $*$-homomorphism that makes the above diagram commutative. $\square$

The pair $\bigl(C^*(j(\mathcal{S})), j\bigr)$ appearing in the proof does not depend on the particular choice of the maximal and isometric map j, as different choices for j produce "equivalent" pairs. To make things precise, one can actually give the "abstract" definition that $C^*_{env}(\mathcal{S}) \simeq (\mathcal{C}, j)$ consists of a completely isometric injection j in a C^*-algebra $\mathcal{C} = C^*(j(\mathcal{C}))$ satisfying the following: if $(\mathcal{C}', j')$ is any other such pair then there exists $*$-homomorphism $\pi : \mathcal{C}' \to \mathcal{C}$ so that $j = \pi \circ j'$. Two such pairs are said to be equivalent provided that the $*$-homomorphism π appearing above is actually isomorphism.

Having defined the C^*-envelope of an operator space abstractly now note that Theorem 2.5.9 implies that if $\mathcal{S}$ is an operator space and $\mathcal{J}$ the Shilov ideal of $\mathcal{S}$ in $C^*(\mathcal{S})$, then $C^*_{env}(\mathcal{S}) \simeq \bigl(C^*(\mathcal{S})/\mathcal{J}, q\bigr)$ is another manifestation of the C^*-envelope as it is conjugate to $\bigl(C^*(j(\mathcal{S})), j\bigr)$ via the $*$-isomorphism $\tilde{\theta}$.

The astute reader has already realized that the C^*-envelope is not just the C^*-algebra $C^*_{env}(\mathcal{S})$ but instead the pair $(C^*_{env}(\mathcal{S}), j)$. This is not to be taken lightly: a pair $(\mathcal{C}, j)$, where $\mathcal{C} \simeq C^*_{env}(\mathcal{S})$ and $j : \mathcal{S} \to \mathcal{C} = C^*(j(\mathcal{C}))$ is an isometry, does not automatically qualify for being $C^*_{env}(\mathcal{S})$.

A natural question arises here: does a unital operator space admit enough maximal *and irreducible* representations to capture the norm? One can weaken somewhat this question by asking irreducibility only for the extension of these maps to the C^*-envelope. The (affirmative) answer to this question for separable algebras was obtained by Arveson [1] and more recently by Davidson and Kennedy [16], without the separability assumption.

A final remark: if $C^*_{env}(\mathcal{S}) \simeq (\mathcal{C}, j)$ with $\mathcal{C} \subseteq B(\mathcal{H})$, does it follow that $j : \mathcal{S} \to B(\mathcal{H})$ is a maximal map? The answer perhaps surprisingly is no! As it turns out not every $*$-representation of $\mathcal{C}$ turns out to restrict to a maximal representation of $\mathcal{S}$. See [55] for more on that.

2.5.1 The C*-envelope of an arbitrary operator algebra

If $\mathcal{A} \subseteq B(\mathcal{H})$ is an operator algebra with $I_\mathcal{H} \notin \mathcal{A}$, then $\mathcal{A}_1$ will denote subspace generated by adjoining $I_\mathcal{H}$ to $\mathcal{A}$.

Theorem 2.5.11 (Meyer [52], 2001). *Let $\varphi \colon \mathcal{A} \to B(\mathcal{K})$ be a completely contractive homomorphism and assume that $I_\mathcal{H} \notin \mathcal{A}$. Then its unitization $\varphi_1 \colon \mathcal{A}_1 \to \mathcal{B}(\mathcal{K})$ is also completely contractive.*

Non-selfadjoint operator algebras 57

This result implies that for a non-degenerately acting, non-unital operator algebra $\mathcal{A}$, its unitization is unique: if $j : \mathcal{A} \to B(\mathcal{K})$ is any completely isometric *homomorphism* (perhaps degenerate),then $I_{\mathcal{K}} \neq j(\mathcal{A})$ and applying the above theorem twice we obtain that j_1 is a complete isometry as well. The situation becomes nicer if we restrict our attention to operator algebras with a contractive approximate unit, i.e., approximately unital operator algebras. In that case the absence of a unit for $\mathcal{A}$ is equivalent to the absence of the unit for any C*-cover. The same also holds for non-degeneracy. Therefore we don't have to talk about non-degenerately acting operator algebras as we make this a blanket assumption for the representations of the C*-covers.

We may consider the category of operator algebras with morphisms the completely contractive homomorphisms.

Corollary 2.5.12. *Every cc homomorphism* $\varphi : \mathcal{A} \to B(\mathcal{H})$ *of an operator algebra* $\mathcal{A}$ *can be dilated to a maximal cc homomorphism* $\varphi' : \mathcal{S} \to B(\mathcal{H}')$.

It seems though that for the above corollary to work, we need to allow for perhaps degenerate dilations as it is not clear what happens to a non-degenerate dilation of a unital operator algebra when we remove the unit. This complicates things.

A different approach suggests that the C*-envelope of an approximately unital operator algebra $\mathcal{A}$ is the C*-algebra generated by $\mathcal{A}$ inside $\mathrm{C}^*_{\mathrm{env}}(\mathcal{A}_1)$ with the obvious injection for $\mathcal{A}$. This provides the last prerequisite for reading the proof of the following result. Note that the case where (X, A) is injective and strict was obtained by Muhly and Solel in [55].

Theorem 2.5.13 (Katsoulis and Kribs [46], 2006). *If* (X, A, φ) *is a C*-correspondence, then*

$$(\mathrm{C}^*_{env}(\mathcal{T}^+_X), j) \simeq \mathcal{O}_X$$

via a map j that sends generators to generators.

Proof. I will give a proof only in the case where A has a unit, which by non-degeneracy is a unit for $\mathcal{O}_X$ as well. By Corollary 2.3.15 we can view $\mathcal{T}^+_X$ as a canonical subalgebra of $\mathcal{O}_X$. It is enough to prove that $\mathcal{J}_{\mathcal{T}^+_X} = \{0\}$, where $\mathcal{J}_{\mathcal{T}^+_X}$ denotes the Shilov ideal of $\mathcal{T}^+_X \subseteq \mathcal{O}_X$. By way of contradiction assume otherwise.

Note that $\mathcal{O}_X$ admits a natural gauge action that leaves $\mathcal{A}$ invariant element wise and twists X by unimodular scalars. Hence that gauge action leaves $\mathcal{T}^+_X$ invariant. Since $\mathcal{J}_{\mathcal{T}^+_X}$ is the largest ideal in $\mathcal{O}_X$ so that the corresponding quotient map is completely isometric on $\mathcal{T}^+_X$, we conclude that $\mathcal{J}_{\mathcal{T}^+_X}$ is gauge invariant and so $\mathcal{O}_X / \mathcal{J}_{\mathcal{T}^+_X}$ admits a gauge action. Since the natural quotient map $q : \mathcal{O}_X \to \mathcal{O}_X / \mathcal{J}_{\mathcal{T}^+_X}$ is not faithful, Theorem 2.3.12 implies that q is not faithful on A and therefore on $\mathcal{T}^+_X$. This is a contradiction. $\square$

2.6 Dynamics and classification of operator algebras

An important moment for non-selfadjoint operator algebra theory was the use of ideas from [11] in the work of Gunther Cornelissen and Matilde Marcolli [8]. Cornelissen and Marcolli actually solved a problem in class field theory by making heavy use of operator algebra theory, including the theory presented in this section. Here is an outline of their result.

A complex number a is called *algebraic* if there exists a non-zero polynomial $p(X) \in \mathbb{Q}[X]$ such that $p(a) = 0$. The polynomial is unique if we require that it be irreducible and monic. We say that a is an algebraic integer if the unique irreducible, monic polynomial which it satisfies has integer coefficients. We know that the set $\overline{\mathbb{Q}}$ of all algebraic numbers is a field, and the algebraic integers form a ring. For an algebraic number a, the set K of all $f(a)$, with $f(X) \in \mathbb{Q}[X]$, is a field, called an algebraic number field. If all the roots of the polynomial $p(X)$ are in K, then K is called Galois over $\mathbb{Q}$.

Question. Which invariants of a number field characterize it up to isomorphism?

The absolute Galois group of a number field K is the group $G_K = Gal(\overline{\mathbb{Q}}/K)$ consisting of all automorphisms σ of $\overline{\mathbb{Q}}$ such that $\sigma(a) = a$ for all $a \in K$. Let $f(X) \in K[X]$ be irreducible, and let Z_f be the set of its roots. The group of permutations of Z_f is a finite group, which is given the discrete topology. Then G_K acts on Z_f. We put a topology on G_K, so that the homomorphism of G_K to the group of permutations of Z_f is continuous for every such $f(X)$. Then G_K is a topological group; it is compact and totally disconnected.

Theorem 2.6.1 (Uchida, 1976). *Number fields E and F are isomorphic as fields if and only if G_E and G_F are isomorphic as topological groups.*

The absolute Galois group is not well understood at all (it is considered an anabelian object). What we do understand well are abelian Galois groups. For a number field K we denote by K^{ab} the maximal abelian extension of K. This is the maximal extension which is Galois (i.e., any irreducible polynomial which has a root in K^{ab} has all its roots in it) such that the Galois group of K^{ab} over K is abelian. For example, the theorem of Kronecker and Weber says that $\mathbb{Q}^{ab}$ is the field generated by all the numbers $\exp(\frac{2\pi i}{n})$, i.e., by all roots of unity.

Example 2.6.2. *The abelianized Galois groups of $\mathbb{Q}(\sqrt{-2})$ and $\mathbb{Q}(\sqrt{-3})$ are isomorphic.*

Theorem 2.6.3 (Cornelissen and Marcolli [8]). *Let E and F be number fields. Then, E and F are isomorphic if and only if there exists an isomorphism of topological groups*

$$\psi \colon G_E^{ab} \to G_F^{ab}$$

such that for every character χ of G_F^{ab} we have $L_{F,\chi} = L_{E,\psi \circ \chi}$, where $L_{F,\chi}$ denotes the L-function associated with ψ.

Cornelissen and Marcolli make essential use of my work with Ken Davidson on multivariable dynamics [11]. At the epicenter of this interaction between number theory and non-selfadjoint operator algebras lies the concept of piecewise conjugacy and the fact that piecewise conjugacy is an invariant for isomorphisms between certain operator algebras associated with multivariable dynamical systems.

Recall from Section 2.2 the context of a topological dynamical system (X, σ) where X locally compacts Hausdorff space and $\sigma : X \to X$ is a proper continuous map or its C*-algebraic analogue (A, α) where A is a C*-algebra and $\sigma : A \to A$ non-degenerate *-endomorphism.

We may consider multivariable analogues of the above concepts. A pair (X, σ) is a multivariable dynamical system provided that X is a locally compact Hausdorff and $\sigma = (\sigma_1, \sigma_2, \ldots, \sigma_n)$, where $\sigma_i : X \to X$, $1 \leq i \leq n$, are continuous (proper) maps. A similar definition holds for a multivariable C*-dynamical system (A, α).

We would like to have an operator algebra $\mathcal{A}$ that encodes the dynamics of (X, σ). Therefore $\mathcal{A}$ should contain a copy of $C_0(X)$ and $S_1, \ldots, S_n$ satisfying covariance relations

$$f S_i = S_i (f \circ \sigma_i)$$

for $1 \leq i \leq n$ and $f \in C_0(X)$.

For $w \in \mathbb{F}_n^+$, say $w = i_k \ldots i_1$, we write $S_w = S_{i_k} \ldots S_{i_1}$. The covariance algebra is

$$\mathcal{A}_0 = \left\{ \sum_{w \in \mathbb{F}_n^+} S_w f_w : f_w \in C_0(X) \right\},$$

where $\mathbb{F}_n^+$ is the free semigroup on n letters. This is an algebra since

$$(S_v)(f S_w g) = S_{vw}(f \circ \sigma_w)g,$$

where $\sigma_w \equiv \sigma_{i_k} \circ \cdots \circ \sigma_{i_1}$. We need a norm condition in order to complete $\mathcal{A}_0$. In the multivariable setting we have more than one choice. Two are the most prominent

(1) Isometric: $S_i^* S_i = I$ for $1 \leq i \leq n$.

(2) Row Isometric: $\begin{bmatrix} S_1 & S_2 & \ldots & S_n \end{bmatrix}^* \begin{bmatrix} S_1 & S_2 & \ldots & S_n \end{bmatrix} = I$.

Completing $\mathcal{A}_0$ using (1) yields the semicrossed product $C_0(X) \times_\sigma \mathbb{F}_n^+$, while completing $\mathcal{A}_0$ using (2) yields the tensor algebra $\mathcal{T}_+(X, \sigma)$. (See Definition 2.4.4.)

2.6.1 Piecewise conjugate multisystems

In order to classify our multivariable algebras up to isomorphism, we need a new notion of conjugacy. An obvious one would be to say that two multivariable dynamical systems (X, σ) and (Y, τ) are *conjugate* if there exists a homeomorphism γ of X onto Y and a permutation $\alpha \in S_n$ so that $\tau_i = \gamma \sigma_{\alpha(i)} \gamma^{-1}$ for $1 \leq i \leq n$. This is too strong for our purposes.

Definition 2.6.4 (Davidson and Katsoulis [11], 2011). *We say that two multivariable dynamical systems (X, σ) and (Y, τ) are piecewise conjugate if there is a homeomorphism γ of X onto Y and an open cover $\{\mathcal{U}_\alpha : \alpha \in S_n\}$ of X so that for each $\alpha \in S_n$,*

$$\gamma^{-1} \tau_i \gamma|_{\mathcal{U}_\alpha} = \sigma_{\alpha(i)}|_{\mathcal{U}_\alpha}.$$

The difference between the two concepts of conjugacy lies on the fact that the permutations depend on the particular open set. As we shall see, a single permutation generally will not suffice.

Here are two examples of when piecewise conjugacy implies conjugacy. Both are taken from [11].

Proposition 2.6.5. *Let (X, σ) and (Y, τ) be piecewise conjugate multivariable dynamical systems. Assume that X is connected and that*

$$E := \{x \in X : \sigma_i(x) = \sigma_j(x), \text{for some } i \neq j\}$$

has empty interior. Then (X, σ) and (Y, τ) are conjugate.

For $n = 2$, we can be more definitive.

Proposition 2.6.6. *. Let X be connected and let $\sigma = (\sigma_1, \sigma_2)$; and let E as above. Then piecewise conjugacy coincides with conjugacy if and only if $\overline{X \setminus E}$ is connected.*

2.6.2 The multivariable classification problem

We want to repeat the success of Theorem 2.2.7. As it turns out we only succeed in the "difficult" direction of that theorem: necessity of conjugacy for isomorphism.

Theorem 2.6.7 (Davidson and Katsoulis [11], 2011). *Let (X, σ) and (Y, τ) be two multivariable dynamical systems. If $\mathcal{T}_+(X, \sigma)$ and $\mathcal{T}_+(Y, \tau)$ or $C_0(X) \times_\sigma \mathbb{F}_n^+$ and $C_0(Y) \times_\tau \mathbb{F}_n^+$ are isomorphic as algebras, then the dynamical systems (X, σ) and (Y, τ) are piecewise conjugate.*

For the tensor algebras, sufficiency holds in the following cases:

(i) X has covering dimension 0 or 1.

Non-selfadjoint operator algebras 61

(ii) σ consists of no more than 3 maps. ($n \leq 3$.)

Theorem 2.6.8 (Davidson and Katsoulis [11], 2011). *Suppose that X is a compact subset of $\mathbb{R}$. Then for two multivariable dynamical systems (X, σ) and (Y, τ), the following are equivalent:*

1. (X, σ) *and* (Y, τ) *are piecewise topologically conjugate.*

2. $\mathcal{T}_+(X, \sigma)$ *and* $\mathcal{T}_+(Y, \tau)$ *are isomorphic.*

3. $\mathcal{T}_+(X, \sigma)$ *and* $\mathcal{T}_+(Y, \tau)$ *are completely isometrically isomorphic.*

The analysis of the $n = 3$ case is the most demanding and required nontrivial topological information about the Lie group $SU(3)$. The conjectured converse reduces to a question about the unitary group $U(n)$.

Conjecture 2.6.9. Let Π_n be the $n!$-simplex with vertices indexed by S_n. Then there should be a continuous function u of Π_n into $U(n)$ so that

1. Each vertex is taken to the corresponding permutation matrix,

2. For every pair of partitions (A, B) of the form

$$\{1, \ldots, n\} = A_1 \dot{\cup} \ldots \dot{\cup} A_m = B_1 \dot{\cup} \ldots \dot{\cup} B_m,$$

where $|A_s| = |B_s|$, $1 \leq s \leq m$, let

$$\mathcal{P}(A, B) = \{\alpha \in S_n : \alpha(A_s) = B_s, 1 \leq s \leq m\}.$$

If $x = \sum_{\alpha \in \mathcal{P}(A,B)} x_\alpha \alpha$, then the non-zero matrix coefficients of $u_{ij}(x)$ are supported on $\bigcup_{s=1}^m B_s \times A_s$. We call this the *block decomposition condition*.

We have established this conjecture for $n = 2$ and 3 and Chris Ramsey has verified the cases $n = 4, 5$.

With Ken Davidson we considered only classical dynamical systems (dynamical systems over commutative C*-algebras) and our notion of piecewise conjugacy applies exclusively to such systems. Motivated by the interaction between number theory and non-selfadjoint operator algebras, one wonders whether a useful analogue of piecewise conjugacy can be developed for multivariable systems over arbitrary C*-algebras. The goal here is to obtain a natural notion of piecewise conjugacy that generalizes that of Davidson and Katsoulis from the commutative case while remaining an invariant for isomorphisms between non-selfadjoint operator algebras associated with such systems. This was undertaken successfully with Kakariadis in [41].

Definition 2.6.10. *Let A be a unital C*-algebra and let $P(A)$ be its pure state space equipped with the w^*-topology. The Fell spectrum $\hat{A}$ of A is the space of unitary equivalence classes of non-zero irreducible representations of A. (The usual unitary equivalence of representations will be denoted as $\sim$.) The GNS construction provides a surjection $P(A) \to \hat{A}$ and $\hat{A}$ is given the quotient topology.*

62   *Recent Advances in Operator Theory and Operator Algebras*

Let A be a unital C*-algebra A and $\alpha = (a_1, a_2, \ldots, a_n)$ be a multivariable system consisting of unital *-epimorphisms. Any such system (A, α) induces a multivariable dynamical system $(\hat{A}, \hat{\alpha})$ over its Fell spectrum $\hat{A}$.

Definition 2.6.11. *Two multivariable systems (A, α) and (B, β) are said to be piecewise conjugate on their Fell spectra if the induced systems $(\hat{A}, \hat{\alpha})$ and $(\hat{B}, \hat{\beta})$ are piecewise conjugate, in the sense of the definition above.*

We have the following result with Kakariadis.

Theorem 2.6.12 (Kakariadis and Katsoulis [41], 2014). *Let (A, α) and (B, β) be multivariable dynamical systems consisting of *-epimorphisms. Assume that either $\mathcal{T}_+(A, \alpha)$ and $\mathcal{T}_+(B, \beta)$ or $A \times_\alpha \mathbb{F}_{n_\alpha}^+$ and $B \times_\beta \mathbb{F}_{n_\beta}^+$ are isometrically isomorphic. Then the multivariable systems (A, α) and (B, β) are piecewise conjugate over their Fell spectra.*

Problem. Is there an analogous result for the Jacobson spectrum?

In particular this implies that when the associated operator algebras are isomorphic then both (A, α) and (B, β) have the same number of *-epimorphisms. (We call this property invariance of the dimension.) In the commutative case, the invariance of the dimension holds for systems consisting of arbitrary endomorphisms. Is it true here?

Theorem 2.6.13 (Kakariadis and Katsoulis [41], 2014). *There exist multivariable systems (A, α_1, α_2) and $(B, \beta_1, \beta_2, \beta_3)$ consisting of *-monomorphisms for which $\mathcal{T}_+(A, \alpha_1, \alpha_2)$ and $\mathcal{T}_+(B, \beta_1, \beta_2, \beta_3)$ are isometrically isomorphic.*

Problem (Invariance of dimension for semicrossed products). Let (A, α) and (B, β) be multivariable dynamical systems consisting of *-endomorphisms. Prove or disprove: if $A \times_\alpha \mathbb{F}_{n_\alpha}^+$ and $B \times_\beta \mathbb{F}_{n_\beta}^+$ are isometrically isomorphic then $n_\alpha = n_\beta$.

We say that two multivariable C*-dynamical systems (A, α) and (B, β) are *outer conjugate* if they have the same dimension and there are *-isomorphism $\gamma : A \to B$, unitary operators $U_i \in B$ and $\pi \in S_n$ so that

$$\gamma^{-1} \alpha_i \gamma(b) = U_i^* \beta_{\pi(i)}(b) U_i$$

for all $b \in B$ and i.

Theorem 2.6.14 (Kakariadis and Katsoulis [41], 2014). *Let (A, α) and (B, β) be two automorphic multivariable C*-dynamical systems and assume that A is primitive. Then the following are equivalent:*

(i) $A \times_\alpha \mathbb{F}_{n_\alpha}^+$ and $B \times_\beta \mathbb{F}_{n_\beta}^+$ are isometrically isomorphic.

(ii) $\mathcal{T}^+(A, \alpha)$ and $\mathcal{T}^+(B, \beta)$ are isometrically isomorphic.

(iii) (A, α) and (B, β) are outer conjugate.

Let us sketch the proof that (i) or (ii) implies (iii) in the above theorem. Assume now that (A, α) and (B, β) are two multivariable dynamical systems such that $\mathcal{T}^+(A, \alpha)$ and $\mathcal{T}^+(B, \beta)$ (or $A \times_\alpha \mathbb{F}^+_{n_\alpha}$ and $B \times_\beta \mathbb{F}^+_{n_\beta}$) are isometrically isomorphic via a mapping α. Since α is isometric, it follows that $\alpha|_A$ is a $*$-monomorphism that maps A onto B. (This is the only point where we use that α is isometric.) We will be denoting $\alpha|_A$ by α as well.

Let S_i, $i = 1, \ldots, n_\alpha$ (resp. T_i, $i = 1, 2, \ldots, n_\beta$) be the generators in $\mathcal{T}^+(A, \alpha)$ (resp. $\mathcal{T}^+(B, \beta)$) and let b_{ij} be the T_i-Fourier coefficient of $\alpha(s_j)$, i.e.,

$$\alpha(S_j) = b_{0j} + T_1 b_{1j} + T_2 b_{2j} + \cdots + T_n b_{nj} + Y,$$

where Y involves Fourier terms of order 2 or higher.

Since α is a homomorphism,

$$\alpha(a)\alpha(S_j) = \alpha(aS_j) = \alpha(S_j \alpha_j(a)) = \alpha(S_j)\alpha\alpha_j(a),$$

for all $a \in A$. Hence, $\beta_i \alpha(a) b_{ij} = b_{ij} \alpha\alpha_j(a)$, $a \in A$, and so

$$\beta_i(b) b_{ij} = b_{ij} \alpha\alpha_j \alpha^{-1}(b) = b_{ij} \tilde{\alpha}_j(b)$$

for all $b \in B$.

From the intertwining equation

$$\beta_i(b) b_{ij} = b_{ij} \tilde{\alpha}_j(b), b \in B \qquad (*)$$

we obtain

- Since A is primitive, $b_{i,j}$ is either zero or invertible!

- If $b_{ij} \neq 0$ then $\beta_i \sim \tilde{\alpha}_j$.

Therefore each equivalence class $\{\beta_1, \beta_2, \ldots, \beta_n\}$ is equivalent to exactly one class $\{\tilde{\alpha}_1, \tilde{\alpha}_2, \ldots, \tilde{\alpha}_m\}$. The proof will follow if we show that $m = n$. By way of contradiction assume that $m < n$.

Start with an "arbitrary" n-tuple $(y_1, y_2, \ldots, y_n)$. From the equation

$$T_1 y_1 + T_2 y_2 + \cdots + T_n y_n = \lim_e \alpha(x_e),$$

where x_e are non-commutative polynomials in $S_1, S_2, \ldots, S_m$ and remembering that

$$\alpha(S_j) = b_{0j} + T_1 b_{1j} + T_2 b_{2j} + \cdots + T_n b_{nj} + Y,$$

we obtain

$$y_1 = \lim_e b_{11} x_e^1 + b_{12} x_e^2 + \cdots + b_{1m} x_e^m$$
$$y_2 = \lim_e b_{21} x_e^1 + b_{22} x_e^2 + \cdots + b_{2m} x_e^m$$
$$\vdots$$
$$y_n = \lim_e b_{n1} x_e^1 + b_{n2} x_e^2 + \cdots + b_{nm} x_e^m.$$

Perform Gaussian elimination to reduce this system to

$$\bar{y}_2 = \lim_e \bar{b}_{22} x_e^2 + \bar{b}_{23} x_e^3 + \cdots + \bar{b}_{2m} x_e^m$$
$$\bar{y}_3 = \lim_e \bar{b}_{32} x_e^2 + \bar{b}_{33} x_e^3 + \cdots + \bar{b}_{3m} x_e^m$$
$$\vdots$$
$$\bar{y}_n = \lim_e \bar{b}_{n2} x_e^2 + \bar{b}_{n3} x_e^3 + \cdots + \bar{b}_{nm} x_e^m.$$

We continue this short of "Gaussian elimination" and we arrive at a system that contains one column and at least two non-trivial rows of the form

$$w_1 = \lim_e d_1 x_e^m$$
$$w_2 = \lim_e d_2 x_e^m,$$

where the data (w_1, w_2) are arbitrary. Therefore d_1, d_2 are non-zero, hence invertible. By letting $w_1 = 1$ we obtain that $\lim_e x_e^m = d_1^{-1}$. Therefore, if we let $w_2 = 0$, then we get that $0 = d_2 d_1^{-1}$, which is a contradiction.

The case of tensor algebras in Theorem 2.6.14 is a special case of the following much more general result. (See Definition 2.4.4 for the C*-correspondences appearing below.)

Theorem 2.6.15 (Kakariadis and Katsoulis [41], 2014). *Let (A, α) and (B, β) be multivariable dynamical systems consisting of $*$-epimorphisms. The tensor algebras $\mathcal{T}_+(A, \alpha)$ and $\mathcal{T}_+(B, \beta)$ are isometrically isomorphic if and only if the correspondences (X_α, A) and (X_β, B) are unitarily equivalent.*

In light of the above result we ask the following problem

Problem. Let (A, α) and (B, β) be multivariable dynamical systems consisting of $*$-monomorphisms. If the tensor algebras $\mathcal{T}_+(A, \alpha)$ and $\mathcal{T}_+(B, \beta)$ are isometrically isomorphic does it follow that the correspondences (X_α, A) and (X_β, B) are unitarily equivalent?

One final remark. The results from [41] involving non-commutative multivariable systems apply to the classification of their tensor algebras up to isometric isomorphism. It is natural to ask what happens with bounded isomorphisms. In light of Theorem 2.6.8 and other results from [11], one might expect that these results persist but we have not been able to verify that. Note here that Dor-On [22] has established that in general the classification of tensor algebras of C*-correspondences up to isometric isomorphism differs from their classification up to algebraic isomorphism. Nevertheless it is not clear that this dichotomy applies to [41].

2.7 Crossed products of operator algebras

As we have seen so far, most crossed product-type constructions in the theory of non-selfadjoint operator algebras involve the action of a semigroup which rarely happens to be a group, on an operator algebra which is usually a C^*-algebra. There is a good reason for this and it goes back to the early work of Arveson who recognized that in order to better encode the dynamics of a homeomorphism σ acting on a locally compact space $\mathcal{X}$, one should abandon group actions and instead focus on the action of $\mathbb{Z}^+$ on $C_0(\mathcal{X})$ implemented by the positive iterates of σ. This theme was fully explored in Sections 2.2 and 2.6.

In this section we follow a less-traveled path: we start with an arbitrary operator algebra, preferably non-selfadjoint, and we allow a whole group to act on it. It is remarkable that there have been no systematic attempts to build a comprehensive theory around such algebras even though this class includes all crossed product C^*-algebras. Admittedly, our interest in group actions on non-selfadjoint operator algebras arose reluctantly as well. Indeed, apart from certain important cases, the structure of automorphisms for non-selfadjoint operator algebras is not well understood. Our initial approach stemmed from an attempt to settle two open problems regarding semi-Dirichlet algebras (which we do settle using the crossed product). We soon realized that even for very "simple" automorphisms (gauge actions), the crossed product demonstrates an unwieldy behavior that allows for significant results.

Definition 2.7.1. *Let $(\mathcal{A}, \mathcal{G}, \alpha)$ be a dynamical system and let $(\mathcal{C}, j)$ be a C^*-cover of $\mathcal{A}$. Then $(\mathcal{C}, j)$ is said to be α-admissible, if there exists a continuous group representation $\dot{\alpha} : \mathcal{G} \to \mathrm{Aut}(\mathcal{C})$ which extends the representation*

$$\mathcal{G} \ni s \mapsto j \circ \alpha_s \circ j^{-1} \in \mathrm{Aut}(j(\mathcal{A})). \tag{2.1}$$

Since $\dot{\alpha}$ is uniquely determined by its action on $j(\mathcal{A})$, both (2.1) and its extension $\dot{\alpha}$ will be denoted by the symbol α.

Definition 2.7.2 (Relative Crossed Product). *Let $(\mathcal{A}, \mathcal{G}, \alpha)$ be a dynamical system and let $(\mathcal{C}, j)$ be an α-admissible C^*-cover for $\mathcal{A}$. Then, $\mathcal{A} \rtimes_{\mathcal{C}, j, \alpha} \mathcal{G}$ and $\mathcal{A} \rtimes^r_{\mathcal{C}, j, \alpha} \mathcal{G}$ will denote the subalgebras of the crossed product C^*-algebras $\mathcal{C} \rtimes_\alpha \mathcal{G}$ and $\mathcal{C} \rtimes^r_\alpha \mathcal{G}$, respectively, which are generated by $C_c(\mathcal{G}, j(\mathcal{A})) \subseteq C_c(\mathcal{G}, \mathcal{C})$.*

One has to be a bit careful with Definition 2.7.2 when dealing with an *abstract* operator algebra. It is common practice in operator algebra theory to denote a C^*-cover by the use of set theoretic inclusion. Nevertheless a C^*-cover for $\mathcal{A}$ is not just an inclusion of the form $A \subseteq \mathcal{C}$ but instead a pair $(\mathcal{C}, j)$, where $\mathcal{C}$ is a C^*-algebra, $j : \mathcal{A} \to \mathcal{C}$ is a complete isometry, and $\mathcal{C} = C^*(j(\mathcal{A}))$. Furthermore, in the case of an α-admissible C^*-cover, it seems

66   *Recent Advances in Operator Theory and Operator Algebras*

that the structure of the relative crossed product for $\mathcal{A}$ should depend on the nature of the embedding j and one should keep that in mind when working with that crossed product. To put it differently, assume that $(\mathcal{A}, \mathcal{G}, \alpha)$ is a dynamical system and $(\mathcal{C}_i, j_i)$, $i = 1, 2$, are C*-covers for $\mathcal{A}$. Further assume that the representations $\mathcal{G} \ni s \mapsto j_i \circ \alpha_s \circ j_i^{-1} \in \mathrm{Aut}(j_i(\mathcal{A}))$ extend to *-representations $\alpha_i : \mathcal{G} \to \mathrm{Aut}(\mathcal{C}_i)$, $i = 1, 2$. It is not at all obvious that whenever $\mathcal{C}_1 \simeq \mathcal{C}_2$ (or even $\mathcal{C}_1 = \mathcal{C}_2$), the C*- dynamical systems $(\mathcal{C}_i, \mathcal{G}, \alpha_i)$ are conjugate nor that the corresponding crossed product algebras are isomorphic. Therefore the (admittedly) heavy notation $\mathcal{A} \rtimes_{\mathcal{C}, j, \alpha} \mathcal{G}$ and $\mathcal{A} \rtimes_{\mathcal{C}, j, \alpha}^r \mathcal{G}$ seems to be unavoidable. Nevertheless, whenever there is no source of confusion, we opt for the lighter notation $\mathcal{A} \rtimes_{\mathcal{C}, \alpha} \mathcal{G}$ and $\mathcal{A} \rtimes_{\mathcal{C}, \alpha}^r \mathcal{G}$. For instance, this is the case when the C*-covers involved are coming either from the C*-envelope or from the universal C*-algebra of $\mathcal{A}$, as the following result shows.

Lemma 2.7.3. *Let $(\mathcal{A}, \mathcal{G}, \alpha)$ be a dynamical system and let $(\mathcal{C}_i, j_i)$ be C*-covers for $\mathcal{A}$ with either $\mathcal{C}_i \simeq \mathrm{C}_{env}^*(\mathcal{A})$, $i = 1, 2$, or $\mathcal{C}_i \simeq \mathrm{C}_{max}^*(\mathcal{A})$, $i = 1, 2$. Then there exist continuous group representations $\alpha_i : \mathcal{G} \to \mathrm{Aut}(\mathcal{C}_i)$ which extend the representations*

$$\mathcal{G} \ni s \mapsto j_i \circ \alpha_s \circ j_i^{-1} \in \mathrm{Aut}(j_i(\mathcal{A})), \quad i = 1, 2.$$

Furthermore $\mathcal{A} \rtimes_{\mathcal{C}_1, j_1, \alpha_1} \mathcal{G} \simeq \mathcal{A} \rtimes_{\mathcal{C}_2, j_2, \alpha_2} \mathcal{G}$ and $\mathcal{A} \rtimes_{\mathcal{C}_1, j_1, \alpha_1}^r \mathcal{G} \simeq \mathcal{A} \rtimes_{\mathcal{C}_2, j_2, \alpha_2}^r \mathcal{G}$, via complete isometries that map generators to generators.

Definition 2.7.4 (Full Crossed Product)**.** *If $(\mathcal{A}, \mathcal{G}, \alpha)$ is a dynamical system then*

$$\mathcal{A} \rtimes_\alpha \mathcal{G} \equiv \mathcal{A} \rtimes_{\mathrm{C}_{max}^*(\mathcal{A}), \alpha} \mathcal{G}.$$

In the case where $\mathcal{A}$ is a C*-algebra then $\mathcal{A} \rtimes_\alpha \mathcal{G}$ is nothing else but the full crossed product C*-algebra of $(\mathcal{A}, \mathcal{G}, \alpha)$. In the general case of an operator algebra, one might be tempted to say that $\mathcal{A} \rtimes_\alpha \mathcal{G} \simeq \mathcal{A} \rtimes_{\mathrm{C}_{env}^*(\mathcal{A}), \alpha} \mathcal{G}$. This is not so clear. First, it is not true in general that $\mathrm{C}_{max}^*(\mathcal{A}) \simeq \mathrm{C}_{env}^*(\mathcal{A})$ and as it turns out, $\mathrm{C}_{max}^*(\mathcal{A})$ is a much more difficult object to identify than $\mathrm{C}_{env}^*(\mathcal{A})$. Furthermore, any covariant representation of $(\mathrm{C}_{env}^*(\mathcal{A}), \mathcal{G}, \alpha)$ extends some covariant representation of $(\mathcal{A}, \mathcal{G}, \alpha)$. The problem is that the converse may not be true, i.e., a covariant representation of $(\mathcal{A}, \mathcal{G}, \alpha)$ does not necessarilly extend to a covariant representation of $(\mathrm{C}_{env}^*(\mathcal{A}), \mathcal{G}, \alpha)$. The identification $\mathcal{A} \rtimes_\alpha \mathcal{G} \simeq \mathcal{A} \rtimes_{\mathrm{C}_{env}^*(\mathcal{A}), \alpha} \mathcal{G}$ is a major open problem, which is resolved only in the case where $\mathcal{G}$ is amenable or when $\mathcal{A}$ is Dirichlet.

For the moment let us characterize the crossed product as the universal object for covariant representations of the dynamical system $(\mathcal{A}, \mathcal{G}, \alpha)$. In the case where $\mathcal{A}$ is a C*-algebra, this was done by Raeburn in [66]. In the generality appearing below, this result is new.

Theorem 2.7.5 (Katsoulis and Ramsey [6])**.** *Let $(\mathcal{A}, \mathcal{G}, \alpha)$ be a dynamical system. Assume that $\mathcal{B}$ is an approximately unital operator algebra such that*

    (i) there exists a completely isometric covariant representation $(j_\mathcal{A}, j_\mathcal{G})$ of $(\mathcal{A}, \mathcal{G}, \alpha)$ into $M(\mathcal{B})$,

Non-selfadjoint operator algebras 67

(ii) given a covariant representation $(\pi, u, \mathcal{H})$ of $(\mathcal{A}, \mathcal{G}, \alpha)$, there is a completely contractive, non-degenerate representation $L : \mathcal{B} \to B(\mathcal{H})$ such that $\pi = \bar{L} \circ j_\mathcal{A}$ and $u = \bar{L} \circ j_\mathcal{G}$, and

(iii) $\mathcal{B} = \overline{\mathrm{span}}\{j_\mathcal{A}(a)\tilde{\jmath}_\mathcal{G}(z) \mid a \in \mathcal{A}, z \in C_c(\mathcal{G})\}$,

where

$$\tilde{\jmath}_\mathcal{G}(z) \equiv \int_\mathcal{G} z(s) j_\mathcal{G}(s) d\mu(s), \quad \text{for all } z \in C_c(\mathcal{G}).$$

Then there exists a completely isometric isomorphism $\rho : \mathcal{B} \to \mathcal{A} \rtimes_\alpha \mathcal{G}$ such that

$$\bar{\rho} \circ j_\mathcal{A} = i_\mathcal{A} \text{ and } \bar{\rho} \circ j_\mathcal{G} = i_\mathcal{G}, \tag{2.2}$$

where $(i_\mathcal{A}, i_\mathcal{G})$ is the canonical covariant representation of $(\mathcal{A}, \mathcal{G}, \alpha)$ in $M(\mathcal{A} \rtimes_\alpha \mathcal{G})$.

In the case where $\mathcal{G}$ is amenable, all relative full crossed products coincide as the next result shows. Its proof requires an essential use of the theory of maximal dilations, as presented in Section 2.5.

Theorem 2.7.6 (Katsoulis and Ramsey [6]). *Let $(\mathcal{A}, \mathcal{G}, \alpha)$ be a dynamical system with $\mathcal{G}$ amenable and let $(\mathcal{C}, j)$ be an α-admissible C*-cover for $\mathcal{A}$. Then*

$$\mathcal{A} \rtimes_\alpha \mathcal{G} \simeq \mathcal{A} \rtimes_{\mathcal{C}, j, \alpha} \mathcal{G} \simeq \mathcal{A} \rtimes^r_{\mathcal{C}, j, \alpha} \mathcal{G}$$

via a complete isometry that maps generators to generators.

One of the central problems of our theory is whether or not the identity

$$C^*_{\mathrm{env}}(\mathcal{A} \rtimes_\alpha \mathcal{G}) = C^*_{\mathrm{env}}(\mathcal{A}) \rtimes_\alpha \mathcal{G} \tag{2.3}$$

is valid. Fortunately in the case where $\mathcal{G}$ is an abelian group we show that the above identity is indeed valid. The case where $\mathcal{G}$ is discrete follows easily from the work we have done so far and from the ideas of either [39] in the $\mathbb{Z}$ case or more directly from [10, Theorem 3.3], by choosing $P = \mathcal{G}$, $\tilde{\alpha} = \alpha$ and transposing the covariance relations. In the generality appearing below, the result is new and paves the way for exploring non-selfadjoint versions of Takai duality.

Theorem 2.7.7 (Katsoulis and Ramsey [6]). *Let $(\mathcal{A}, \mathcal{G}, \alpha)$ be a unital dynamical system and assume that $\mathcal{G}$ is an abelian locally compact group. Then*

$$C^*_{env}(\mathcal{A} \rtimes_\alpha \mathcal{G}) \simeq C^*_{env}(\mathcal{A}) \rtimes_\alpha \mathcal{G}.$$

Here is the promised version of Takai duality for *arbitrary* operator algebras. We will make shortly an important use of that duality in our investigation for the semisimplicity of crossed products.

Let $(\mathcal{A}, \mathcal{G}, \alpha)$ be a dynamical system with $\mathcal{G}$ an abelian locally compact group. Let $\hat{\mathcal{G}}$ be the Pontryagin dual of $\mathcal{G}$. The dual action $\hat{\alpha}$ is defined on $C_c(\mathcal{G}, \mathcal{A})$ by $\hat{\alpha}_\gamma(f)(s) = \overline{\gamma(s)} f(s)$, $f \in C_c(\mathcal{G}, \mathcal{A})$, $\gamma \in \hat{\mathcal{G}}$.

Theorem 2.7.8 (Takai duality, Katsoulis and Ramsey [6]). *Let* $(\mathcal{A}, \mathcal{G}, \alpha)$ *be a dynamical system with* $\mathcal{G}$ *a locally compact abelian group. Then*

$$\left(\mathcal{A} \rtimes_\alpha \mathcal{G}\right) \rtimes_{\hat{\alpha}} \hat{\mathcal{G}} \simeq \mathcal{A} \otimes \mathcal{K}\left(L^2(\mathcal{G})\right),$$

where $\mathcal{K}\left(L^2(\mathcal{G})\right)$ *denotes the compact operators on* $L^2(\mathcal{G})$ *and* $\mathcal{A} \otimes \mathcal{K}\left(L^2(\mathcal{G})\right)$ *is the subalgebra of* $\mathrm{C}^*_{env}(\mathcal{A}) \otimes \mathcal{K}\left(L^2(\mathcal{G})\right)$ *generated by the appropriate elementary tensors.*

Let us give an application of our theory to solve a problem that actually motivated our investigation. In [12] Davidson and Katsoulis introduced the class of semi-Dirichlet algebras. The semi-Dirichlet property is a property satisfied by all tensor algebras and the premise of [12] is that this is the actual property that allows for such a successful dilation and representation theory for the tensor algebras. Indeed in [12] the authors verified that claim by recasting many of the tensor algebra results in the generality of semi-Dirichlet algebras. What was not clear in [12] was whether there exist "natural" examples of semi-Dirichlet algebras beyond the classes of tensor and Dirichlet algebras. It turns out that the crossed product is the right tool for generating new examples of semi-Dirichlet algebras from old ones. By also gaining a good understanding on Dirichlet algebras and their crossed products we were able to answer in [6] a related question of Ken Davidson: we produced the first examples of semi-Dirichlet algebras which are neither Dirichlet algebras nor tensor algebras (Corollary 2.7.9).

Corollary 2.7.9. *There exist semi-Dirichlet algebras which are neither Dirichlet nor isometrically isomorphic to the tensor algebra of any* C^*-*correspondence.*

Recall the definition of the Jacobson Radical of a (not necessarily unital) ring.

Definition 2.7.10. *Let* $\mathcal{R}$ *be a ring. The Jacobson radical* $\mathrm{Rad}\,\mathcal{R}$ *is defined as the intersection of all maximal regular right ideals of* $\mathcal{R}$*. (A right ideal* $\mathfrak{I} \subseteq \mathcal{R}$ *is regular if there exists* $e \in \mathcal{R}$ *such that* $ex - x \in \mathfrak{I}$, *for all* $x \in \mathcal{R}$.)

An element x in a ring R is called right quasi-regular if there exists $y \in \mathcal{R}$ such that $x + y + xy = 0$. It can be shown that $x \in \mathrm{Rad}\,\mathcal{R}$ if and only if xy is right quasi-regular for all $y \in \mathcal{R}$. This is the same as $1 + xy$ being right invertible in $\mathcal{R}^1$ for all $y \in \mathcal{R}$.

In the case where $\mathcal{R}$ is a Banach algebra we have

$$\mathrm{Rad}\,\mathcal{R} = \{x \in \mathcal{R} \mid \lim_n \|(xy)^n\|^{1/n} = 0, \text{ for all } y \in \mathcal{R}\}$$

$$= \{x \in \mathcal{R} \mid \lim_n \|(yx)^n\|^{1/n} = 0, \text{ for all } y \in \mathcal{R}\}.$$

A ring $\mathcal{R}$ is called semisimple iff $\mathrm{Rad}\,\mathcal{R} = \{0\}$.

The study of the various radicals is a central topic of investigation in

Non-selfadjoint operator algebras

Abstract Algebra and Banach Algebra theory. In Operator Algebras, the Jacobson radical and the semisimplicity of operator algebras have been under investigation since the very beginning of the theory.

Our next result uncovers a new permanence property in the theory of crossed products.

Theorem 2.7.11 (Katsoulis and Ramsey [6]). *Let $(\mathcal{A}, \mathcal{G}, \alpha)$ be a dynamical system with $\mathcal{G}$ a discrete abelian group. If $\mathcal{A}$ is semisimple then $\mathcal{A} \rtimes_\alpha \mathcal{G}$ is semisimple.*

Proof. Assume that the crossed product is not semisimple and so there is a non-zero $a \in \operatorname{Rad} \mathcal{A} \rtimes_\alpha \mathcal{G}$. Any isometric automorphism fixes the Jacobson radical and so $\Phi_g(a) = a_g \in \operatorname{Rad} \mathcal{A} \rtimes_\alpha \mathcal{G}$ for all $g \in \mathcal{G}$, where $a \sim \sum_{g \in G} a_g U_g$. By a standard result in operator algebra theory involving the Fejer kernel, since $a \neq 0$ there is a $g \in \mathcal{G}$ such that $a_g \neq 0$. This implies that $a_g b$ is quasinilpotent for all $b \in \mathcal{A}$ and so $a_g \in \operatorname{Rad} \mathcal{A}$. Therefore, $\mathcal{A}$ is not semisimple. $\square$

The previous result raises two natural questions. Is the converse of Theorem 2.7.11 true? Is Theorem 2.7.11 valid beyond discrete abelian groups? As we shall see shortly, both questions have a negative answer. To see this for the first question, we investigate a class of operator algebras which was quite popular in the 1990s, the triangular AF algebras [15, 18, 19, 20, 29, 50, 65].

Definition 2.7.12. *Let $\mathcal{A}$ be a strongly maximal TAF algebra. The dynamical system $(\mathcal{A}, \mathcal{G}, \alpha)$ is said to be linking if for every matrix unit $e \in \mathcal{A}$ there exists a group element $g \in G$ such that $e\mathcal{A}\alpha_g(e) \neq \{0\}$.*

By Donsig's criterion if $\mathcal{A}$ is semisimple then $(\mathcal{A}, \mathcal{G}, \alpha)$ is linking. The following example shows that there are other linking dynamical systems.

Example 2.7.13. *Let $\mathcal{A}_n = \mathbb{C} \oplus \mathcal{T}_{2n}$ and define the embeddings $\rho_n : \mathcal{A}_n \to \mathcal{A}_{n+1}$ by*

$$\rho_n(x \oplus a) = x \oplus \begin{bmatrix} x & & \\ & a & \\ & & x \end{bmatrix}.$$

Then $\mathcal{A} = \varinjlim \mathcal{A}_n$ is a strongly maximal TAF algebra that is not semisimple. Consider the following map $\psi : \mathcal{A}_n \to \mathcal{A}_{n+1}$ given by

$$\psi(x \oplus a) = x \oplus \begin{bmatrix} x & & \\ & x & \\ & & a \end{bmatrix}.$$

You can see that $\psi \circ \rho_n = \rho_{n+1} \circ \psi$ on $\mathcal{A}_n$ and so ψ is a well-defined map on $\cup \mathcal{A}_n$. By considering that

$$\psi^{-1}(x \oplus a) = x \oplus \begin{bmatrix} a & & \\ & x & \\ & & x \end{bmatrix}$$

one gets $\psi \circ \psi^{-1} = \psi^{-1} \circ \psi = \rho_{n+1} \circ \rho_n$ on $\mathcal{A}_n$. Hence, ψ extends to be an isometric automorphism of $\mathcal{A}$. Finally, for every $e_{i,j}^{2n} \in \mathcal{A}_n, i \neq j$,

$$
e_{i,j}^{(2n)} \begin{bmatrix} 0_{2n} & & \\ & 0_{2n} & e_{j,i}^{(2n)} \\ & & 0_{2n} \end{bmatrix} \psi^{(2n)}(e_{i,j}^{(2n)})
$$

$$
= \begin{bmatrix} 0_{2n} & & \\ & e_{i,j}^{(2n)} & \\ & & 0_{2n} \end{bmatrix} \begin{bmatrix} 0_{2n} & & \\ & 0_{2n} & e_{j,i}^{(2n)} \\ & & 0_{2n} \end{bmatrix} \begin{bmatrix} 0_{2n} & & \\ & 0_{2n} & \\ & & e_{i,j}^{(2n)} \end{bmatrix}
$$

$$
= \begin{bmatrix} 0_{2n} & & \\ & 0_{2n} & e_{i,j}^{(2n)} \\ & & 0_{2n} \end{bmatrix}.
$$

Therefore, $(\mathcal{A}, \mathbb{Z}, \psi)$ is a linking dynamical system.

The following theorem and the previous example establish that the converse of Theorem 2.7.11 is not true in general.

Theorem 2.7.14 (Katsoulis and Ramsey [6]). *Let $\mathcal{A}$ be a strongly maximal TAF algebra and $\mathcal{G}$ a discrete abelian group. The dynamical system $(\mathcal{A}, \mathcal{G}, \alpha)$ is linking if and only if $\mathcal{A} \rtimes_\alpha G$ is semisimple.*

In order to answer the other question we need the following.

Lemma 2.7.15. *Let $\mathcal{A}$ be an operator algebra and let $\mathcal{K}(\mathcal{H})$ denote the compact operators acting on a separable Hilbert space $\mathcal{H}$. If $\mathcal{A} \otimes \mathcal{K}(\mathcal{H})$ is semisimple, then $\mathcal{A}$ is semisimple.*

Proof. Identify $\mathcal{A} \otimes \mathcal{K}(\mathcal{H})$ with the set of all infinite operator matrices $[(a_{ij})]_{i,j=1}^\infty$ with entries in $\mathcal{A}$, which satisfy

$$
\left\| [(a_{ij})]_{i,j=1}^\infty - [(a_{ij})]_{i,j=1}^m \right\| \xrightarrow[m \to \infty]{} 0.
$$

By way of contradiction, assume that $0 \neq x \in \operatorname{Rad} \mathcal{A}$. Let

$$
X = x \otimes e_{11} \in A \otimes \mathcal{K}(\mathcal{H})
$$

be the infinite operator matrix whose $(1,1)$-entry is equal to x and all other entries are 0.

If $A = [(a_{ij})]_{i,j=1}^\infty \in A \otimes \mathcal{K}(\mathcal{H})$, then an easy calculation shows that

$$
(AX)^n = \begin{pmatrix} (a_{11}x)^n & 0 & 0 & \cdots \\ a_{21}x(a_{11}x)^{n-1} & 0 & 0 & \cdots \\ a_{31}x(a_{11}x)^{n-1} & 0 & 0 & \cdots \\ \vdots & & \vdots & \vdots & \ddots \end{pmatrix}
$$

$$
= A\big((a_{11}x)^{n-1} \otimes e_{11}\big).
$$

Hence

$$\lim_n \|(AX)^n\|^{1/n} \le \lim_n \|A\|^{1/n} \limsup_n \|(a_{11}x)^{n-1}\|^{1/n}$$

$$= \limsup_n \|(a_{11}x)^n\|^{1/n} = 0$$

because $x \in \mathrm{Rad}\,\mathcal{A}$. Hence $0 \ne X \in \mathrm{Rad}\,\mathcal{A} \otimes \mathcal{K}(\mathcal{H})$, which is the desired contradiction. $\qquad\square$

We now show that Theorem 2.7.11 does not necessarily hold for groups which are not discrete and abelian. Using our Takai duality, we can actually show that this fails even for $\mathbb{T}$.

Example 2.7.16. *A dynamical system $(\mathcal{B}, \mathbb{T}, \beta)$, with $\mathcal{B}$ a semisimple operator algebra, for which $\mathcal{B} \rtimes_\beta \mathbb{T}$ is not semisimple.*

We will employ again our previous results and Takai duality. In Example 2.7.13 we saw a linking dynamical system $(\mathcal{A}, \mathbb{Z}, \alpha)$ for which $\mathcal{A}$ is not semisimple. Since $(\mathcal{A}, \mathbb{Z}, \alpha)$ is linking, we have by Theorem 2.7.14 that the algebra $\mathcal{B} \equiv \mathcal{A} \rtimes_\alpha \mathbb{Z}$ is semisimple. Let $\beta \equiv \hat{\alpha}$. Then,

$$\mathcal{B} \rtimes_\beta \mathbb{T} = \big(\mathcal{A} \rtimes_\alpha \mathbb{Z}\big) \rtimes_{\hat{\alpha}} \mathbb{T} \simeq \mathcal{A} \rtimes \mathcal{K}\big(\ell^2(\mathbb{Z})\big),$$

which is not semisimple.

Quite interestingly, the converse of Theorem 2.7.11 holds for compact abelian groups. Once again, the result follows from Takai duality.

Theorem 2.7.17 (Katsoulis and Ramsey [6]). *Let $(\mathcal{A}, \mathcal{G}, \alpha)$ be a dynamical system, with $\mathcal{G}$ a compact, second countable abelian group. If $\mathcal{A} \rtimes_\alpha \mathcal{G}$ is semisimple, then $\mathcal{A}$ is semisimple.*

Proof. Assume that $\mathcal{A} \rtimes_\alpha \mathcal{G}$ is semisimple. Then Theorem 2.7.11 implies that $\big(\mathcal{A} \rtimes_\alpha \mathcal{G}\big) \rtimes_{\hat{\alpha}} \hat{\mathcal{G}}$ is semisimple. By Takai duality, $\mathcal{A} \otimes \mathcal{K}\big(L^2(\mathcal{G})\big)$ is semisimple and so by Lemma 2.7.15, $\mathcal{A}$ is semisimple, as desired. $\qquad\square$

Another natural question in the theory of crossed products asks whether or not the class of tensor algebras is being preserved under crossed products by a locally compact abelian group. Our next result shows that this is not the case.

Theorem 2.7.18 (Katsoulis and Ramsey [6]). *Let $\mathcal{G}$ be a discrete amenable group and let $\alpha : \mathcal{G} \to \mathrm{Aut}\big(\mathbb{A}(\mathbb{D})\big)$ be a representation. Assume that the common fixed points of the Möbius transformations associated with $\{\alpha_g\}_{g\in\mathcal{G}}$ do not form a singleton. Then $\mathbb{A}(\mathbb{D}) \rtimes_\alpha \mathcal{G}$ is a Dirichlet algebra which is not isometrically isomorphic to the tensor algebra of any C^*-correspondence.*

Nevertheless for a special class of dynamical systems we obtain a positive answer.

Theorem 2.7.19 (Katsoulis and Ramsey [6]). *Let $\mathcal{A}$ be a tensor algebra and let $\alpha : \mathcal{G} \to \operatorname{Aut} \mathcal{A}$ be the action of a locally compact group $\mathcal{G}$ by gauge actions. Then the relative crossed product $\mathcal{A} \rtimes_{C^*_{env}(\mathcal{A}),\alpha} \mathcal{G}$ is the tensor algebra of a C^*-correspondence.*

It turns out that the above result allows us to reformulate a problem in C^*-algebra theory, the Hao–Ng Isomorphism Conjecture, into a problem concerning the C^*-envelope of a crossed product operator algebra and the validity of (2.3). We direct the reader to [6] for more details and to [42] for a positive resolution of the Hao–Ng problem for all discrete groups. It is perhaps surprising that a seemingly selfadjoint problem, such as the Hao–Ng isomorphism problem, requires the use of non-selfadjoint operator algebras for its solution.

2.8 Local maps and representation theory

The study of local maps, i.e., local derivation, local multipliers etc, has a long history in operator algebras [31, 32, 33, 34, 35, 36, 37]. My involvement with such maps started somewhat accidentally. I was trying to understand whether or not a compact adjointable operator on a Hilbert C^*-module has a non-trivial invariant subspace. Of course it does and as it turns out the adjointable operators have lots of *common* invariant subspaces provided the C^*-module is not a Hilbert space. Characterizing these common invariant subspaces and other peripheral results however required a result of Barry Johnson on local multipliers [32], which sparked my interest on the topic.

Definition 2.8.1. *If $\mathcal{X}$ is Banach space and $\mathcal{S} \subseteq B(\mathcal{X})$, then $\mathcal{S}$ is said to be reflexive iff the following condition is satisfied*

$$T \in B(\mathcal{X}),\ Tx \in [\mathcal{S}x],\ \forall x \in \mathcal{X} \implies T \in \mathcal{S}.$$

For $\mathcal{S} \subseteq X$ unital algebra this is equivalent to the familiar

$$T(M) \subseteq M, \forall M \in \operatorname{Lat} \mathcal{S} \implies T \in \mathcal{S}.$$

Theorem 2.8.2 (Katsoulis [44], 2014). *Let E a Hilbert C^*-module over a C^*-algebra $\mathcal{A}$ and let $L(E)$ be the adjointable operators on E. Then*

$$\operatorname{Lat} L(E) = \{ E\mathcal{J} \mid \mathcal{J} \subseteq \overline{\langle E, E \rangle} \text{ closed left ideal } \},$$

where

$$E\mathcal{J} = \{ \xi j \mid \xi \in E, j \in \mathcal{J} \}.$$

and the association $\mathcal{J} \mapsto E\mathcal{J}$ establishes a complete lattice isomorphism between the closed left ideals of $\overline{\langle E, E \rangle}$ and $\operatorname{Lat} L(E)$.

Non-selfadjoint operator algebras

Note that if $\operatorname{End}_A(E)$ denotes the bounded A-module operators on E, then the above implies

$$\operatorname{Lat} L(E) = \operatorname{Lat} \operatorname{End}_A(E).$$

Theorem 2.8.3 (Katsoulis [44], 2014). *Let E be a Hilbert module over a C^*-algebra A. Then*

$$\operatorname{Alg} \operatorname{Lat} \mathcal{L}(E) = \operatorname{End}_A(E).$$

In particular, $\operatorname{End}_A(E)$ is a reflexive algebra of operators acting on E.

The proof follows from the following.

Theorem 2.8.4 (Johnson [32], 1968). *Let A be a semisimple Banach algebra and let Φ be a linear operator acting on A that leaves invariant all closed left ideals of A. Then*

$$\Phi(ba) = \Phi(b)a, \ \forall\, a, b \in A,$$

i.e, Φ is a left multiplier.

In particular, if $1 \in A$ is a unit then Φ is the left multiplication operator by $\Phi(1)$.

Definition 2.8.5. *A map $S : A \to X$ into a right Banach A-module is said to be a local left multiplier iff for every $A \in A$, there exists left multiplier $\Phi_a \in LM(A, X)$ so that $S(a) = \Phi_a(a)$.*

Approximately local multipliers are defined to satisfy an approximate version of the above definition.

Proposition 2.8.6. *Let A be a Banach algebra with approximate unit. If $S \in B(A)$, then the following are equivalent.*

(i) S is a closed left ideal preserver.

(ii) S is an approximately local left multiplier.

(iii) $S \in \operatorname{Lat} \operatorname{Alg} LM(A)$.

Proof. Assume that (i) is valid and let $a \in A$. Note that the set

$$\mathcal{J}_a \equiv \overline{\{ba \mid b \in A\}}^{\|\cdot\|} \subseteq A$$

is a closed left ideal. Hence $S(a) \in \mathcal{J}_a$ and so

$$S(a) = \lim b_n a = \lim_n L_{b_n}(a),$$

i.e., S is an approximate left multiplier.

Assume now that (ii) is true. To show that $S \in \operatorname{Lat} \operatorname{Alg} LM(A)$, it is enough to prove that $S(\mathcal{J}_a) \subseteq \mathcal{J}_a$, for all $a \in A$. However, since S is an approximately local left multiplier

$$S(ba) = \lim_n L_{c_n}(ba) = \lim_n c_n ba \in \mathcal{J}_a$$

and since it is bounded, we obtain (ii).

The rest of the proof follows from similar arguments. $\qquad\square$

74   *Recent Advances in Operator Theory and Operator Algebras*

The proposition above allows us to reformulate Johnson's theorem as follows.

Theorem 2.8.7 (Johnson [32], 1968). *The space $LM(\mathcal{A})$ of left multipliers over a semisimple Banach algebra is reflexive, i.e., $\operatorname{Alg} \operatorname{Lat} M(\mathcal{A}) = \mathcal{A}$.*

Question. What about Johnson's theorem in the context of non-semisimple operator algebras? Is the space $LM(\mathcal{A})$ of left multipliers over an operator algebra reflexive?

It is easy to produce a negative answer even for finite-dimensional algebras.

Example 2.8.8. *If*

$$\mathfrak{A} = \left\{ \begin{pmatrix} \lambda & \mu \\ 0 & \lambda \end{pmatrix} \mid \lambda, \mu \in \mathbb{C} \right\} = \{\lambda I + \mu e_{12} \mid \lambda, \mu \in \mathbb{C}\}$$

then

$$S_{\mathfrak{A}} : \mathfrak{A} \longrightarrow \mathfrak{A} : \lambda I + \mu e_{12} \longmapsto \lambda I + 2\mu e_{12}$$

is a local multiplier which is not a multiplier.

Indeed if $\lambda \neq 0$ then $S_{\mathfrak{A}}(\lambda I + \mu e_{12}) = (I + (\mu/\lambda)e_{12})(\lambda I + \mu e_{12})$ or otherwise $S_{\mathfrak{A}}(\mu e_{12}) = 2I(\mu e_{12})$. This shows that $S_{\mathfrak{A}}$ is a local multiplier. It is easy to see that in the case $\lambda \neq 0$, the factor $(I + \mu/\lambda e_{12})$ is uniquely determined by $\lambda I + \mu e_{12}$ and so $S_{\mathfrak{A}}$ cannot be a multiplier.

Proposition 2.8.9. (Hadwin [33, 34] 1990s). *Let $\mathcal{A}$ be a Banach algebra generated by its idempotents and $\mathcal{X}$ be a right Banach $\mathcal{A}$-module. Then any approximately local left multiplier from $\mathcal{A}$ into $\mathcal{X}$ is a multiplier. Hence $LM(\mathcal{A}, \mathcal{X})$ is reflexive.*

Proof. Let $S : \mathcal{A} \to \mathcal{X}$ be an approximate left multiplier. Note that for any $a, p \in \mathcal{A}$ with $p = p^2$, we have $S(ap) \in \overline{\mathcal{X}p}$ and $S(a(1 - p)) \in \overline{\mathcal{X}(1 - p)}$. Therefore

$$S(a)p = S(ap)p + S(a(1 - p))p$$
$$= S(ap)p = S(ap).$$

Repeated applications of the above establish the result in the case where p is a product of idempotents. Since $\mathcal{A}$ is generated by such, S is a left multiplier, as desired. $\qquad\square$

The previous result formed the basis for a variety of results by Hadwin, Li, Pan Dong, and others to establish reflexivity for $LM(A)$, where $\mathcal{A}$ ranges over a variety of algebras rich in idempotents, including nest and CSL algebras.

What about semicrossed products? Or tensor algebras of multivariable systems? What about their spaces of derivations? Are they reflexive? Are local derivations actually derivations? Such algebras might contain no idempotents. We use instead the Representation theory.

Non-selfadjoint operator algebras

Theorem 2.8.10 (Katsoulis [43]). *If $\mathcal{G} = (\mathcal{G}^0, \mathcal{G}^1, r, s)$ is a topological graph, then the finite-dimensional nest representations of its tensor algebra $\mathcal{T}_{\mathcal{G}}^+$ separate points.*

The above generalizes an earlier result of Davidson and Katsoulis (2006) regarding tensor algebras of graphs.

Corollary 2.8.11. *If $\mathcal{G} = (\mathcal{G}^0, \mathcal{G}^1, r, s)$ is a topological graph, then $LM(\mathcal{T}_{\mathcal{G}}^+)$ is reflexive.*

Proof. Let S be an approximately local multiplier on $\mathcal{T}_{\mathcal{G}}^+$ and let

$$\rho_i : \mathcal{T}_{\mathcal{G}}^+ \to B(\mathcal{H}_i), \quad i \in \mathbb{I}$$

separating family of representations on finite-dimensional Hilbert space so that each $\rho_i(\mathcal{T}_{\mathcal{G}}^+)$ is a finite-dimensional nest algebra. Furthermore, $\rho_i(\mathcal{T}_{\mathcal{G}}^+)$ is a right $\mathcal{T}_{\mathcal{G}}^+ / \ker \rho_i$-module, with the right action coming from ρ_i. Since S preserves closed left ideals we obtain

$$S_i : \mathcal{T}_{\mathcal{G}}^+ / \ker \rho_i \longrightarrow \rho_i(\mathcal{T}_{\mathcal{G}}^+); a + \ker \rho_i \longmapsto \rho_i(S(a)).$$

However S_i is an approximate left multiplier and so Hadwin's theorem implies that S_i is actually a left multiplier. Hence

$$S_i(ab + \ker \rho_i) = S_i(a + \ker \rho_i)\rho_i(b)$$

and so

$$\rho_i\big(S(ab) - S(a)b\big) = 0, \quad \text{for all } i \in \mathbb{I}.$$

Since $\cap_i \ker \rho_i = \{0\}$, the conclusion follows. $\square$

Corollary 2.8.12. *A local left multiplier on $C_0(X) \times_\sigma \mathbb{Z}^+$ is actually a multiplier.*

Problem. Is the same true for $\mathcal{A} \times_\sigma \mathbb{Z}^+$ in the case where $\mathcal{A}$ is a non-commutative C*-algebra?

Remark 2.8.13. *The above Corollary is not valid for multipliers taking values in a $C(X) \times_\sigma \mathbb{Z}^+$-module.*

Theorem 2.8.14 (Katsoulis [43]). *Let $\mathcal{G} = (\mathcal{G}^0, \mathcal{G}^1, r, s)$ be a topological graph and let $\{\mathcal{G}_v\}_{v \in \mathcal{G}^0}$ be the family of discrete graphs associated with $\mathcal{G}$. Assume that the set of all points $v \in \mathcal{G}^0$, for which $\mathcal{G}_v$ is either acyclic or transitive, is dense in $\mathcal{G}^0$. Then any approximately local derivation on $\mathcal{T}_{\mathcal{G}}^+$ is a derivation.*

Corollary 2.8.15 (Katsoulis [43]). *Let (X, σ) be a dynamical system for which the eventual periodic points have empty interior, e.g., σ is a homeomorphism. Then any local derivation on $C_0(X) \times_\sigma \mathbb{Z}^+$ is actually a derivation.*

Problem. What about the case where σ is an arbitrary selfmap?

References

[1] W. Arveson, *The non-commutative Choquet boundary*, J. Amer. Math. Soc. **21** (2008), 1065–1084.

[2] W. Arveson, *Subalgebras of* C*-*algebras*, Acta Math. **123** (1969), 141–224.

[3] W. Arveson, *Operator algebras and measure preserving automorphisms*, Acta Math. **118** (1967), 95–109.

[4] W. Arveson and K. Josephson, *Operator algebras and measure preserving automorphisms II*, J. Funct. Anal. **4** (1969), 100–134.

[5] T. Bates, D. Pask, I. Raeburn, and W. Szymanski, *The* C*-*algebras of row-finite graphs*, New York J. Math. **6** (2000), 307–324.

[6] E. Bedos, S. Kaliszewski, J. Quigg, and D. Robertson, *A new look at crossed product correspondences and associated* C*–*algebras*, J. Math. Anal. Appl. **426** (2015), 1080–1098.

[7] D. Blecher and C. Le Merdy, *Operator algebras and their modules–an operator space approach*, London Mathematical Society Monographs New Series **30**, Oxford University Press, 2004.

[8] G. Cornelissen and M. Marcolli, *Quantum Statistical. Mechanics, L-Series and Anabelian Geometry*, manuscript arXiv:1009.0736.

[9] J. Cuntz, *K-theory for certain* C*-*algebras. II*, J. Operator Theory **5** (1981), 101–108.

[10] K. Davidson, A. Fuller and E.T.A. Kakariadis, *Semicrossed products of operator algebras by semigroups*, Mem. Amer. Math. Soc., in press.

[11] K. Davidson and E. Katsoulis, *Operator algebras for multivariable dynamics*, Mem. Amer. Math. Soc. **209** (2011), no. 982, viii+53 pp.

[12] K. Davidson and E. Katsoulis, *Dilation theory, commutant lifting and semicrossed products*, Doc. Math. **16** (2011), 781–868.

[13] K. Davidson and E. Katsoulis, *Semicrossed products of simple C*-algebras*, Math. Ann. **342** (2008), 515–525.

[14] K. Davidson and E. Katsoulis, *Isomorphisms between topological conjugacy algebras*, J. Reine Angew. Math. **621** (2008), 29–51.

References

[15] K. Davidson and E. Katsoulis, *Primitive limit algebras and* C*-*envelopes*, Adv. Math. **170** (2002), 181–205.

[16] K. Davidson and M. Kennedy, *The Choquet boundary of an operator system*, Duke Math. J. **164** (2015), 2989–3004.

[17] K. Davidson and D. Pitts, *The algebraic structure of non-commutative analytic Toeplitz algebras*, Math. Ann. **311** (1998), 275–303.

[18] A. Donsig, *Semisimple triangular AF algebras*, J. Funct. Anal. **111** (1993), 323–349.

[19] A. Donsig and A. Hopenwasser, *Analytic partial crossed products*, Houston J. Math. **31** (2005),495–527.

[20] A. Donsig and A. Hopenwasser, *Order preservation in limit algebras*, J. Funct. Anal. **133** (1995), 342–394.

[21] A. Donsig, A. Katavolos, and A. Manoussos, *The Jacobson radical for analytic crossed products*, J. Funct. Anal. **187** (2001), 129–145.

[22] A. Dor-On, *Isomorphisms of tensor algebras arising from weighted partial systems*, Trans. Amer. Math. Soc., to appear.

[23] M. Dritschel and S. McCullough, *Boundary representations for families of representations of operator algebras and spaces*, J. Operator Theory **53** (2005), 159–167.

[24] N. Fowler, P. Muhly, and I. Raeburn, *Representations of Cuntz–Pimsner algebras*, Indiana Univ. Math. J. **52** (2003), 569–605.

[25] E. Gootman and A. Lazar, *Crossed products of type I AF* C*-*algebras by abelian groups*, Isr. J. Math **56** (1986), 267–279.

[26] M. Hamana, *Injective envelopes of operator systems*, Publ. Res. Inst. Math. Sci. **15** (1979), 773–785.

[27] D. Hadwin and T. Hoover, *Operator algebras and the conjugacy of transformations*, J. Funct. Anal. **77** (1988), 112–122.

[28] G. Hao and C. K. Ng, *Crossed products of* C*-*correspondences by amenable group actions*, J. Math. Anal. Appl. **345** (2008), 702–707.

[29] T. Hudson, *Ideals in triangular AF algebras*, Proc. London Math. Soc. **69** (1994), 345–376.

[30] S. Itoh, *Conditional expectations in* C*-*crossed products*, Trans. Amer. Math. Soc. **267** (1981), 661–667.

[31] B.E. Johnson, *Local derivations on* C*-*algebras are derivations*, Trans. Amer. Math. Soc. **353** (2001), 313–325.

References 79

[32] B.E. Johnson, *Centralisers and operators reduced by maximal ideals*, J. London Math. Soc. **43** (1968), 231–233.

[33] The Hadwin Lunch Bunch, *Local multiplications on algebras spanned by idempotents*, Linear and Multilinear Algebra **37** (1994), 259–263.

[34] D. Hadwin and J. Kerr, *Local multiplications on algebras*, J. Pure Appl. Algebra **115** (1997), 231–239.

[35] D. Hadwin and J. Li, *Local derivations and local automorphisms on some algebras*, J. Operator Theory **60** (2008), 29–44.

[36] D. Hadwin and J. Li, *Local derivations and local automorphisms*, J. Math. Anal. Appl. **290** (2004), 702–714.

[37] R. Kadison, *Local derivations*, J. Algebra **130** (1990), 494–509.

[38] E.T.A. Kakariadis, *The Dirichlet property for tensor algebras*, Bull. Lond. Math. Soc. **45** (2013), 1119–1130.

[39] E.T.A. Kakariadis and E. Katsoulis, *Semicrossed products of operator algebras and their* C*-*envelopes*, J. Funct. Anal. **262** (2012), 3108–3124.

[40] E.T.A. Kakariadis and E. Katsoulis, *Contributions to* C*-*correspondences...*, Trans. Amer. Math. Soc. **364** (2012), 6605–6630.

[41] E.T.A. Kakariadis and E. Katsoulis, *Isomorphism invariants for multi-variable C*-dynamics*, J. NonCommutative Geometry **8** (2014), 771–787.

[42] E. Katsoulis, C*-*envelopes and the Hao–Ng isomorphism for discrete groups*, Int. Math. Res. Not. IMRN, to appear.

[43] E. Katsoulis, *Local maps and the representation theory of operator algebras*, Trans. Amer. Math. Soc. **368** (2016), 5377–5397.

[44] E. Katsoulis, *The reflexive closure of the adjointable operators*, Illinois J. Math. **58** (2014), 359–367.

[45] E. Katsoulis and D. Kribs, *Isomorphisms of algebras associated with directed graphs*, Math. Ann. **330** (2004), 709–728.

[46] E. Katsoulis and D. Kribs, *Tensor algebras of C*-correspondences and their* C*-*envelopes*, J. Funct. Anal. **234** (2006), 226–233.

[47] E. Katsoulis and C. Ramsey, *Crossed products of operator algebras*, Mem. Amer. Math. Soc. in press , Arxiv 1512.08162v2.

[48] T. Katsura, *On* C*-*algebras associated with* C*-*correspondences*, J. Funct. Anal. **217** (2004), 366–401.

References

[49] T. Katsura, *A class of* C*-*algebras generalizing both graph algebras and homeomorphism* C*-*algebras. I. Fundamental results*, Trans. Amer. Math. Soc. **356** (2004), 4287–4322.

[50] D. Larson and B. Solel, *Structured triangular limit algebras*, Proc. London Math. Soc. **75** (1997), 177–193.

[51] M. McAsey and P. Muhly, *Representations of nonselfadjoint crossed products*, Proc. London Math. Soc. **47** (1983), 128–144.

[52] R. Meyer, *Adjoining a unit to an operator algebra*, J. Operator Theory **46** (2001), 281–288.

[53] P. Muhly, *Radicals, crossed products, and flows*, Ann. Polon. Math. **43** (1983), 35–42.

[54] P. Muhly and B. Solel, *An algebraic characterization of boundary representations*, Nonselfadjoint Operator Algebras, Operator Theory, and Related Topics, Birkhuser Verlag, Basel, 1998, pp. 189–196.

[55] P. Muhly and B. Solel, *Tensor algebras over C*-*correspondences: representations, dilations, and C*-*envelopes*, J. Funct. Anal. **158** (1998), 389–457.

[56] P. Muhly and M. Tomforde, *Adding tails to* C*-*correspondences*, Documenta Mathematica, **9** (2004), 79–106

[57] V. Paulsen, *Completely Bounded Maps and Operator Algebras*, Cambridge Studies in Advanced Mathematics **78**, Cambridge University Press, 2002.

[58] J. Peters, *Semicrossed products of* C*-*algebras*, J. Funct. Anal. **59** (1984), 498–534.

[59] J. Peters, *The ideal structure of certain nonselfadjoint operator algebras*, Trans. Amer. Math. Soc. **305** (1988), 333–352.

[60] M. Pimsner, *A class of* C*-*algebras generalizing both Cuntz–Krieger algebras and crossed products by* $\mathbb{Z}$, Free probability theory (Waterloo, ON, 1995), 189–212, Fields Inst. Commun., 12, Amer. Math. Soc., Providence, RI, 1997.

[61] Y. Poon and B. Wagner, $\mathbb{Z}$-*analytic TAF algebras and dynamical systems*, Houston J. Math. **19** (1993), 181–199.

[62] G. Popescu, *Non-commutative disc algebras and their representations*, Proc. Amer. Math. Soc. **124** (1996), 2137–2148.

[63] G. Popescu, *Free holomorphic automorphisms of the unit ball of* $B(H)^n$, J. Reine Angew. Math. **638** (2010), 119–168.

References

[64] S.C. Power, *Classification of analytic crossed product algebras*, Bull. London Math. Soc. **24** (1992), 368–372.

[65] S.C. Power, *Limit algebras: an introduction to subalgebras of* C*-*algebras*, Pitman Research Notes in Mathematics Series, **278**, Longman Scientific & Technical, Harlow 1992.

[66] I. Raeburn, *On crossed products and Takai duality*, Proc. Edinburgh Math. Soc. **31** (1988), 321–330.

[67] J. Ringrose, *On some algebras of operators*, Proc. London Math. Soc. **15** (1965), 61–83.

[68] D. Williams, *Crossed products of* C*-*algebras*, Mathematical Surveys and Monographs, Vol. 134, American Mathematical Society, 2007.

Chapter 3

An introduction to sofic entropy

David Kerr

3.1	Introduction	83
3.2	Internal and external approximation	85
3.3	Amenable measure entropy	89
3.4	Amenable topological entropy	95
3.5	Sofic measure entropy	98
3.6	Sofic topological entropy	104
3.7	Dualizing sofic measure entropy	107
3.8	Algebraic actions	109
3.9	Further developments	111

3.1 Introduction

Since its introduction into ergodic theory by Kolmogorov in the late 1950s, entropy has played a central role in the study of dynamical systems, with wide-ranging applications to fields such as number theory, smooth dynamics, and operator algebras. One of the highlights of the theory is the isomorphism theorem of Ornstein that produced an entropy classification of Bernoulli shifts, later extended by Ornstein and Weiss to all countably infinite amenable-acting groups.

While Ornstein and Weiss were able to extend much of the abstract entropy theory to the amenable setting in the 1970s and 1980s, they pointed out that, for the free group on two generators, the Bernoulli action with uniform 2-atom base factors onto the Bernoulli action with uniform 4-atom base, there was a phenomenon which is precluded for amenable groups because of the fact that entropy cannot increase undertaking factors [45]. This appeared to spell trouble for any attempt to extend the entropy theory beyond the amenable case. However, in a remarkable breakthrough in the late 2000s Bowen succeeded in formulating a notion of measure entropy for the much broader class of sofic groups [5], and this has led, among other things, to an entropy classification for Bernoulli actions of a large class of nonamenable groups [7]. The basic theory of sofic entropy was further consolidated by Kerr and Li, who

84   *Recent Advances in Operator Theory and Operator Algebras*

extended Bowen's original definition to the generator-free case, introduced sofic topological entropy, and established a variational principle relating the two [30, 33]. In these notes we will provide an introduction to and overview of sofic entropy as it has evolved over the last eight years on the basis of this work of Bowen and of Kerr and Li. Additional details and proofs of many of the results can be found in the book [36].

Section 3.2 provides some additional context and motivation for readers who have some familiarity with operator algebras, where the ideas of external and internal approximation that account for the conceptual distinction between amenability and soficity have long played an instrumental role in elucidating structure and establishing classification theorems. In Section 3.3 we review the Kolmogorov–Sinai approach to entropy in its structurally most general setting of amenable groups, and in Section 3.4 we discuss its topological counterpart. Sections 3.5 and 3.6 introduce the basic theory of sofic measure entropy and sofic topological entropy, respectively. In Section 3.7 we discuss a dual approach to sofic measure entropy which brings it into closer technical alignment with sofic topological entropy. In Section 3.8 we present a recent theorem of Hayes which gives a formula for the sofic entropy of principal algebraic actions in terms of the Fuglede–Kadison determinant. Finally, in Section 3.9 we discuss some further developments.

We round out this introduction with some words on terminology and notation. Throughout these notes G will denote a countable discrete group, with identity element written e. We are primarily interested in the case that G is infinite, but we will not make this a blanket assumption. An action $G \curvearrowright X$ of G on a set X is a map $G \times X \to X$, expressed using the concatenation $(s, x) \mapsto sx$, which satisfies $(st)x = s(tx)$ (thus making the notation stx unambiguous) and $ex = x$ for all $s, t \in G$ and $x \in X$. If we need to give the action a name, say α, we will write $G \overset{\alpha}{\curvearrowright} X$. If A is a subset of X and $s \in G$ then we write sA for the set $\{sx : x \in A\}$. A set $A \subseteq X$ is G-*invariant* if $sA \subseteq A$ for all $s \in G$, in which case we actually have $sA = A$ for all $s \in G$ due to the existence of inverses in G. If $G \curvearrowright X$ and $G \curvearrowright Y$ are two actions of the same group, then a map $\varphi : X \to Y$ is said to be G-*equivariant* if $\varphi(sx) = s\varphi(x)$ for all $s \in G$ and $x \in X$.

We will be working in the following two basic settings of ergodic theory and topological dynamics:

1. X is a standard probability space (X, μ) and the action, which we will write as $G \curvearrowright (X, \mu)$, is measure-preserving, i.e., $\mu(sA) = \mu(A)$ for all measurable sets $A \subseteq X$ (this is what we will refer to as a *p.m.p.* [*probability-measure-preserving*] *action*);

2. X is a compact metrizable space and the action is continuous.

We are thus working in the "finite" realm both measure-theoretically (finite measure, normalized so that X has measure one) and topologically (compact space). It is the tension between the infiniteness of the group and the "finiteness" of the space that furnishes the conditions for the kind of mixing behavior that governs an asymptotic invariant like entropy.

In the occasional instance that we will need to explicitly refer to the σ-algebra of a probability space (X, μ), we will write it as $\mathscr{B}_X$.

Two actions $G \curvearrowright X$ and $G \curvearrowright Y$ of the same group on sets X and Y are said to be *conjugate* if there is a bijection $\varphi : X \to Y$ satisfying $\varphi(sx) = s\varphi(x)$ for all $s \in G$ and $x \in X$ (equivariance). If X and Y are topological spaces and the actions are continuous, then we will understand conjugacy to mean that the map φ can be chosen to be a homeomorphism. If the actions are p.m.p. then we say that they are *measure conjugate* if there is an equivariant measure isomorphism φ between invariant conull sets of X and Y (by *measure isomorphism* we mean an isomorphism between measurable spaces which pushes one measure forward to the other). If we weaken the requirement that φ be a measurable space isomorphism to mere measurability, then we say that the action on Y is a *factor* of the action on X.

Partitions of a probability space are always assumed to be measurable, i.e., their members are measurable sets. The *join* $\mathscr{P}_1 \vee \cdots \vee \mathscr{P}_n$ of finitely many partitions of a probability space is the partition

$$\{A_1 \cap \cdots \cap A_n : A_i \in \mathscr{P}_i \text{ for } i = 1, \ldots, n\}.$$

A partition $\mathscr{Q}$ is a *refinement* of another partition $\mathscr{P}$, written $\mathscr{Q} \geq \mathscr{P}$, if every member of $\mathscr{Q}$ is a union of members of $\mathscr{P}$.

The *join* $\mathscr{U}_1 \vee \cdots \vee \mathscr{U}_n$ of finitely many open covers $\mathscr{U}_1, \ldots, \mathscr{U}_n$ of a topological space is the open cover

$$\{U_1 \cap \cdots \cap U_n : U_i \in \mathscr{U}_i \text{ for } i = 1, \ldots, n\}.$$

An open cover $\mathscr{V}$ is a *refinement* of another open cover $\mathscr{U}$, written $\mathscr{V} \geq \mathscr{U}$, if every member of $\mathscr{V}$ is contained in a member of $\mathscr{U}$.

Logarithms will always be to base e. We denote by $\mathrm{Sym}(A)$ the permutation group of a set A, and write $\mathrm{Sym}(d)$ in the special case $A = \{1, \ldots, d\}$. We write $\mathbb{P}_d$ for the power set of $\{1, \ldots, d\}$ and typically view this as an algebra under the Boolean operations.

Acknowledgments. These notes are based on a lecture series delivered at the workshop "Recent Advances in Operator Theory and Operator Algebras" at the ISI Bangalore in 2014. I thank the organizers for their generous hospitality. Partial support was provided by NSF grants DMS-1162309 and DMS-1500593.

3.2   Internal and external approximation

The subject of functional analysis revolves in large part around the idea of finite or finite-dimensional approximation, whether as a constitutive ingre-

86   *Recent Advances in Operator Theory and Operator Algebras*

dient in some underlying theory, as illustrated by Lebesgue integration in its use of simple functions, or as a property enjoyed by some objects in a category but not others, as exemplified by amenability in its various incarnations for groups, measured equivalence relations, and operator algebras. The concept of amenability, introduced by von Neumann in the late 1920s, is distinguished by the myriad of ways in which it can be expressed, among which is the averaging property that gave birth to its name. It is its formulation as an approximation property, however, that has led to a remarkable structure theory in each of its principal domains of application, as one sees in the Ornstein–Weiss quasitiling machinery for amenable groups, the Connes–Feldman–Weiss theorem for amenable measured equivalence relations, the work of Connes and Haagerup on the classification of amenable (i.e., injective/hyperfinite) von Neumann algebras, and the Elliott classification program for simple separable amenable (i.e., nuclear) C*-algebras.

That a comprehensive structure theory is possible in all of these settings is explained by the fact that the approximation properties characterizing amenability are essentially *internal* in nature and thus in many situations can be used to asymptotically assemble a global picture of the object out of finite-dimensional pieces. In the world of von Neumann algebras with separable predual, for example, amenability can be equivalently expressed, via the work of Connes, as injectivity (extendibility of completely positive maps into the algebra), semidiscreteness (a finite-dimensional approximation property involving completely positive maps), or hyperfiniteness (the algebra can be written as the weak operator closure of an increasing union of finite-dimensional subalgebras). The situation for C*-algebras is not as stark, as it is complicated by the fact that amenability is a measure-theoretic property and hence can mesh in subtle ways with higher-dimensional topological phenomena, which can make their presence felt even in simple C*-algebras but are entirely absent in von Neumann algebras. While C*-algebraic amenability (usually defined as a cohomological property involving derivations, as for von Neumann algebras) is equivalent to a completely positive approximation property analogous to semidiscreteness, as well as to the tensor product norm property of nuclearity, the C*-algebra analogues of hyperfinite von Neumann algebras, known as AF (approximately finite-dimensional) algebras, are very far from exhausting all nuclear C*-algebras, and it is within an intermediate ground between approximately finite-dimensionality and nuclearity, namely finite nuclear dimension, that classifiability by K-theoretic data is achievable, as shown in the remarkable recent work of Elliott–Gong–Lin–Niu [18, 21] and Tikuisis–White–Winter [49] that caps many years of progress in the subject.

In operator algebras there have long been *external* counterparts to the internal finite-dimensional approximation of amenability and its topological/C*-algebraic relatives like approximate finite-dimensionality. Here the arrows get reversed, so that instead of studying maps from finite-dimensional algebras F into our given algebra A we consider maps from A into finite-dimensional algebras F. If we happen to be in the special case where our maps are homo-

An introduction to sofic entropy 87

morphisms then the difference can be effectively and succinctly described as one of (finite-dimensional) subalgebras versus (finite-dimensional) quotients.

The most venerable external approximation property in C*-algebra theory, and certainly the most important in recent years within the nuclear realm, is that of quasidiagonality. In the form due to Voiculescu, this asks, for a C*-algebra A, that there exists a net of completely positive contractions $\varphi_k : A \to M_{n_k}$ into matrix algebras which are asymptotically multiplicative and isometric in the sense that

1. $\|\varphi_k(ab) - \varphi_k(a)\varphi_k(b)\| \to 0$ for all $a, b \in A$, and

2. $\|\varphi_k(a)\| \to \|a\|$ for all $a \in A$.

It has turned out, within the context of the classification program for simple separable nuclear C*-algebras, that quasidiagonality is the key to unlocking certain kinds of fundamental structural information in their most general form, not only in the Toms–Winter conjecture relating $\mathscr{Z}$-stability, finite nuclear dimension, and strict comparison [42], but also in the effort to pinpoint the barest possible hypotheses for K-theoretic classification arguments to be viable in the stably finite case [18, 21].

In fact, in the classification theory for simple separable nuclear C*-algebras it has long been the interplay between the internal and the external that has driven the subject forward. A similar interplay also grounds the classification theory of injective von Neumann algebras, as is clear for example in the discussion of Popa's local quantization and Connes's uniqueness theorem in Sections 11.4 and 11.5 of [10]. The very expression of nuclearity as the completely positive approximation property, and the directly analogous property of semidiscreteness for von Neumann algebras, actually incorporate both viewpoints. For a C*-algebra A, this characterization of nuclearity asks for a net of matrix algebras M_{n_k} and completely positive contractions $\varphi_k : A \to M_{n_k}$ and $\psi_k : M_{n_k} \to A$ which pointwise asymptotically factorize the identity map on A, i.e., $\|(\psi_k \circ \varphi_k)(a) - a\| \to 0$ for all $a \in A$. Semidiscreteness is defined similarly but using trace norm approximations, and the fact that it is equivalent to the more purely internal property of hyperfiniteness is special to the measure-theoretic context of von Neumann algebras, reflecting the purely internal characterization of amenability for groups in terms of the Følner condition.

While C*-algebra classification methods have become extremely sophisticated since the early days of Elliott's pioneering work on AF algebras, they continue to be based on the same kind of intertwining argument, which always involves both an existence and a uniqueness theorem. The idea is to locally lift an isomorphism between the invariants of two algebras to a map between the algebras themselves (existence) and then show that, up to inner automorphisms, there is essentially only one way to do this (uniqueness). One then repeats this in a back-and-forth process, alternating between the external (existence) and internal (uniqueness), in order to asymptotically construct an isomorphism between the two algebras that induces the isomorphism between the invariants. As we will discuss below, amenability for groups is characterized by a closely analogous type of uniqueness when viewed from the external standpoint of sofic approximation, and this duality of perspective has also been advantageous in the world of groups and group actions, although it has only been developed there much more recently.

Quasidiagonality is a strange hybrid of measure (the use of completely positive maps) and topology (norm approximation) that has no combinatorial or geometric counterpart for groups, although it is now known thanks to Tikuisis, White, and Winter [49] that, for a discrete group G, the reduced group C*-algebra $C^*_\lambda(G)$ is quasidiagonal if and only if G is amenable. Dropping the complete positivity and the contractivity yields the purely topological notion of an MF algebra [3], which no longer implies amenability. Indeed Haagerup and Thorbjørnsen showed that the reduced group C*-algebra of a free group on two or more generators belongs in this class [26]. In fact it remains an open problem whether $C^*_\lambda(G)$ is an MF algebra for every discrete group G, and more generally whether every stably finite C*-algebra is MF.

On the von Neumann algebra side is the corresponding property of embeddability into the ultrapower R^ω of the hyperfinite II_1 factor, and it is similarly an open problem whether every II_1 factor satisfies this embeddability (*Connes's embedding problem*), which can be alternatively formulated as finite-dimensional modeling property in the spirit of quasidiagonality and MFness. This question was raised by Connes in the context of his work on injectivity and hyperfiniteness, where he pointed out that the von Neumann algebras of free groups are R^ω-embeddable and that other II_1 factors ought to be as well. The specialization of Connes's embedding problem to groups asks whether every discrete group admits local unitary matrix models up to tracial approximation (*hyperlinearity*), which is equivalent to R^ω-embeddability of the group von Neumann algebra. Soficity is precisely what one obtains if one requires these matrix models to be not merely unitary but in fact permutation matrices. Curiously this property was not formulated until much more recently in the work of Gromov [25], with the terminology being coined and the basic theory consolidated by Weiss in [50]. On the C*-algebra side the relationship between hyperlinearity and soficity translates as that between MF and LEF (*locally embeddable into finite groups*), with the latter meaning that the local approximation in the permutation modeling holds not only in trace but in norm, and hence with zero error, so that we can simply dispense with permutations and ask abstractly for local models in abstract finite groups. Here we see that, in contrast to the measurable domain, when topological properties of C*-algebras are converted via combinatorial analogy to the group setting they become purely group-theoretic. One can complete this picture with one last mix-and-match (internal/group-theoretic) to produce the notion of a locally finite group, in which every finite set generates a finite subgroup.

Among the various finite approximation properties of soficity (external/ measure-theoretic), amenability (internal/measure-theoretic), LEF (external/

group-theoretic), and locally finite (internal/group-theoretic) that we thus obtain through these conceptual pairings, soficity is the most general. The class of sofic groups includes amenable groups as well as residually finite groups (among which count the free groups), and in fact it is not known whether nonsofic groups exist. While soficity does not provide for the kind of detailed structure theory that can be consistently leveraged toward definitive results (e.g., ergodic theorems) as in the amenable setting, its great flexibility turned out to be well suited for defining a dynamical invariant like entropy, as was discovered by Bowen in his seminal work, which greatly expanded the theory of dynamical entropy beyond its classical development. As mentioned in the introduction, that such a generalized entropy theory could be realized was especially surprising given a prior observation of Ornstein and Weiss that such a notion of entropy, if it applied to free groups, would necessarily fail the regularity property of monotonicity under quotients (an instance of an automatic consequence of amenability being replaced by a more complicated and difficult-to-theorize set of possible behaviors).

3.3 Amenable measure entropy

There are two technically distinct but related ways in which the notion of entropy can be made mathematically precise as a measure of the uncertainty of a system whose constituents may exist in one of finitely many states. The first was proposed by Boltzmann in his work on statistical mechanics in the late 1800s and the second by Shannon in his work on communication theory in the 1940s. In a curious reversal of the historical relation between amenability (introduced by von Neumann in the 1920s) and soficity (introduced by Gromov in the 1990s), Boltzmann's entropy is an *external* concept and naturally forms the basis for sofic entropy, as we will see in Section 3.5, while Shannon's entropy is based on an *internal* heuristic and thus is naturally compatible with amenability for the purpose of defining dynamical entropy, as originally shown by Kolmogorov in the late 1950s for integer actions with a generating partition, and then without the generator assumption by Sinai shortly afterwards. We will review here the Kolmogorov–Sinai approach to entropy in the general setting of amenable groups, where it was initiated by Kieffer [37] and then developed by Ornstein and Weiss [45]. We will thus begin by defining Shannon entropy, while our discussion of Boltzmann entropy will be postponed to Section 3.5 where it is needed for sofic entropy.

Let (X, μ) be a probability space and $\mathscr{P} = \{A_1, \ldots, A_n\}$ a finite partition of X. Suppose that x is a randomly chosen point of X, and all that we know about x is that it is contained in a specified member of $\mathscr{P}$, say A_i. We wish to assign a measure of the amount of information that we gain upon learning that the otherwise completely unknown point x belongs to A_i. The smaller the measure of A_i is, the more likely we will be able to distinguish x from another random point y whose membership in one of the sets $A_1, \ldots, A_n$ is similarly all the information we know concerning it, and so it is reasonable to assign $\mu(A_i)^{-1}$ as the information gain. However, as we wish to express the average information over all points in X, it is better to convert this to a quantity that behaves additively, and so we take a logarithm and use $-\log \mu(A_i)$ instead. This gives us the *information function*

$$I_{\mathscr{P}}(x) = -\sum_{i=1}^{n} \mathbf{1}_{A_i} \log \mu(A_i)$$

of the partition $\mathscr{P}$, where $\mathbf{1}_{A_i}$ is the indicator function of the set A_i. The *Shannon entropy* of the partition $\mathscr{P}$ is then defined as the integral of the information function:

$$H(\mathscr{P}) = \int_X I_{\mathscr{P}} \, d\mu(x) = \sum_{i=1}^{n} -\mu(A_i) \log \mu(A_i).$$

This can also be defined more generally for countable partitions, although it will not be necessary for us to do so in these notes (see however the first paragraph of Section 3.9).

One can furthermore conditionalize Shannon entropy with respect to a second partition $\mathscr{Q} = \{B_1, \ldots, B_m\}$, so that it captures the information gain in learning that a random point x belongs to a specified member of $\mathscr{P}$ given that we already know x belongs to a specified member of $\mathscr{Q}$. Knowing that x lies in a particular B_j, we then consider this B_j as the ambient probability space under the normalized measure $A \mapsto \mu(A)/\mu(B_j)$ and revert to the prescription from before for defining the information gain in learning that x belongs to a particular A_j, so that this quantity is now $-\log(\mu(A_i \cap B_j)/\mu(B_j))$. The *conditional information function* is then

$$I_{\mathscr{P},\mathscr{Q}} = -\sum_{j=1}^{m}\sum_{i=1}^{n} \mathbf{1}_{A_i \cap B_j} \log \frac{\mu(A_i \cap B_j)}{\mu(B_j)}$$

and the *conditional (Shannon) entropy* of $\mathscr{P}$ given $\mathscr{Q}$ is

$$H(\mathscr{P}|\mathscr{Q}) = \int_X I_{\mathscr{P},\mathscr{Q}} \, d\mu = \sum_{j=1}^{m}\sum_{i=1}^{n} -\mu(A_i \cap B_j) \log \frac{\mu(A_i \cap B_j)}{\mu(B_j)}.$$

One can push this further by replacing $\mathscr{P}$ with a sub-σ-algebra $\mathscr{C}$, with the conditional information function now given by

$$I_{\mathscr{P},\mathscr{C}} = \sum_{j=1}^{n} (-\log \mathbb{E}_{\mathscr{C}}(\mathbf{1}_{A_j})) \mathbf{1}_{A_j},$$

where $\mathbb{E}_{\mathscr{C}}$ is the conditional expectation with respect to $\mathscr{C}$, and the *conditional entropy* of $\mathscr{P}$ given $\mathscr{C}$ defined as

$$H(\mathscr{P}|\mathscr{C}) = \int_X I_{\mathscr{P},\mathscr{C}} \, d\mu.$$

By associating to $\mathscr{Q}$ the σ-algebra it generates we recover from this the conditional entropy $H(\mathscr{P}|\mathscr{Q})$. Conditional entropy allows one to develop a powerful calculus for carrying out estimates in the entropy theory for actions of amenable groups, especially in averaging arguments. This calculus is grounded in the following basic properties of Shannon entropy. Here $\mathscr{P}$, $\mathscr{Q}$, and $\mathscr{R}$ are finite partitions of X and $T : X \to X$ is a p.m.p. transformation.

1. $0 \le H(\mathscr{P}) \le \log |\mathscr{P}|$,

2. $H(\mathscr{P}) = \log |\mathscr{P}|$ if and only if the elements of $\mathscr{P}$ all have the same measure,

3. if $\mathscr{P} \le \mathscr{Q}$ then $H(\mathscr{P}|\mathscr{R}) \le H(\mathscr{Q}|\mathscr{R})$,

4. if $\mathscr{R} \le \mathscr{Q}$ then $H(\mathscr{P}|\mathscr{Q}) \le H(\mathscr{P}|\mathscr{R})$,

5. $0 \le H(\mathscr{P}|\mathscr{Q}) \le H(\mathscr{P})$,

6. $H(\mathscr{P}|\mathscr{Q}) = H(\mathscr{P})$ if and only if $\mathscr{P}$ and $\mathscr{Q}$ are independent,

7. $H(\mathscr{P}|\mathscr{Q}) = 0$ if and only if $\mathscr{P} \le \mathscr{Q}$ modulo null sets,

8. $H(\mathscr{P} \vee \mathscr{Q}|\mathscr{R}) = H(\mathscr{P}|\mathscr{R}) + H(\mathscr{Q}|\mathscr{P} \vee \mathscr{R})$,

9. $H(T\mathscr{P}|T\mathscr{Q}) = H(\mathscr{P}|\mathscr{Q})$.

Now we bring our countable discrete group G into the picture. In order to carry out the kind of averaging of Shannon entropy that will yield a viable asymptotic invariant, we will need G to be amenable, which means that there is a unital positive linear map (or *mean*) $\sigma : \ell^\infty(G) \to \mathbb{C}$ which is invariant under the action of G on $\ell^\infty(G)$ given by $(sf)(t) = f(s^{-1}t)$ for all $f \in \ell^\infty(G)$ and $s, t \in G$. Amenability can be expressed in a great variety of different ways, but the most consequential, because of its concrete combinatorial character, is the Følner property, which asks that there exists a sequence $\{F_n\}$ of nonempty finite subsets of G which are asymptotically left invariant in the sense that

$$\lim_{n \to \infty} \frac{|sF_n \Delta F_n|}{|F_n|} = 0$$

for all $s \in G$, where Δ denotes symmetric difference. Such a sequence is called a *(left) Følner sequence*. Amenability can be equivalently described via the existence of an asymptotically right invariant sequence (a *right Følner sequence*), and even via the existence of an asymptotically two-sided invariant sequence (a *two-sided Følner sequence*), but for the purpose of developing the

92   *Recent Advances in Operator Theory and Operator Algebras*

entropy theory it is generally sufficient to work with one-sided invariance, which we choose by convention to be from the left. The Følner condition for a sequence $\{F_n\}$ can also be expressed by saying that for every finite set $K \subseteq G$ containing e one has

$$\lim_{n \to \infty} \frac{|\{s \in F_n : Ks \subseteq F_n\}|}{|F_n|} = 1. \tag{3.1}$$

In other words, if n is large then most right translates of K which intersect F_n are in fact fully contained in F_n. This formulation of the Følner condition often turns out to be more useful in applications, as will be illustrated in our discussion of surjunctivity in Section 3.4.

The prototype of an infinite amenable group is $\mathbb{Z}$ and the prototype of a Følner sequence is the sequence of intervals $\{0, \ldots, n-1\}$ in $\mathbb{Z}$. All Abelian groups are amenable, as one can see from the structure theorem for finitely generated Abelian groups. Amenability is moreover preserved undertaking subgroups, quotients, extensions, and increasing unions. One can thus construct many examples by starting with Abelian groups and finite groups and applying any sequence of these operations. This results in the class of *elementary amenable groups*, which is strictly smaller than the class of all amenable groups, as was shown by Grigorchuk. What distinguishes $\mathbb{Z}$ from other amenable groups are both its freeness and the fact that it gives expression to the notions of future and past. This has the effect that many results in the ergodic theory of amenable groups, when formulated for $\mathbb{Z}$, take a particularly strong form and/or admit simpler proofs that do not generalize or even otherwise make sense for other groups (e.g., proofs based on recurrence). It is quite surprising just how much of the entropy theory originally developed for $\mathbb{Z}$ works for arbitrary amenable groups, as one can see most strikingly in the work of Ornstein and Weiss [45].

In the original Kolmogorov–Sinai setting of $\mathbb{Z}$-actions, i.e., a single p.m.p. transformation $T : X \to X$ under iteration, we first define the entropy with respect to a finite partition $\mathscr{P}$ by setting

$$h(T, \mathscr{P}) = \lim_{n \to \infty} \frac{1}{n} H(\mathscr{P} \vee T^{-1}\mathscr{P} \vee \cdots \vee T^{-n-1}\mathscr{P}),$$

where the limit can be shown to exist by a simple exercise using the subadditivity of Shannon entropy recorded as item (8) above. This local quantity can then be turned into a conjugacy invariant for the action by taking the supremum over all finite partitions $\mathscr{P}$, which defines the *(Kolmogorov–Sinai) measure entropy*

$$h(T) = \sup_{\mathscr{P}} h(T, \mathscr{P})$$

of the transformation T. What is crucial in showing that this is a nontrivial and computable invariant is the fact that we are averaging over subsets of $\mathbb{Z}$, namely the intervals $\{0, \ldots, n-1\}$, which become asymptotically invariant under translation as $n \to \infty$. Indeed if we have two finite partitions $\mathscr{P}$ and $\mathscr{Q}$

such that $\mathcal{Q} \leq T^j \mathscr{P} \vee T^{j+1} \mathscr{P} \vee \cdots \vee T^k \mathscr{P}$ for some integers $j < k$, and $n \in \mathbb{N}$ is extremely large compared to $k - j$, then using the properties of Shannon entropy we see that

$$
\begin{aligned}
\frac{1}{n} H(\mathcal{Q} \vee T^{-1} \mathcal{Q} \vee \cdots \vee T^{-n-1} \mathcal{Q}) &\leq \frac{1}{n} H(T^k \mathscr{P} \vee T^{k-1} \mathscr{P} \vee \cdots \vee T^{-n-1+j} \mathscr{P}) \\
&= \frac{1}{n} H(\mathscr{P} \vee T^{-1} \mathscr{P} \vee \cdots \vee T^{-n-1-(k-j)} \mathscr{P}) \\
&\approx \frac{1}{n} H(\mathscr{P} \vee T^{-1} \mathscr{P} \vee \cdots \vee T^{-n-1} \mathscr{P}).
\end{aligned}
$$

It follows that $h(T, \mathcal{Q}) \leq h(T, \mathscr{P})$.

We can moreover reach the same conclusion if $\mathcal{Q}$ is merely contained in the σ-algebra generated modulo null sets by the partitions $T^k \mathscr{P}$ for $k \in \mathbb{Z}$. In this case $\mathcal{Q}$ is approximately refined by partitions of the form $T^j \mathscr{P} \vee T^{j+1} \mathscr{P} \vee \cdots \vee T^k \mathscr{P}$ and hence is approximated by partitions $\mathcal{Q}'$ that are actually refined by such a join, where approximation can be interpreted in any of various natural ways, one of which is to say that both $H(\mathcal{Q}|\mathcal{Q}')$ and $H(\mathcal{Q}'|\mathcal{Q})$ are small. Using the easily verified fact that

$$
h(T, \mathcal{Q}) \leq h(T, \mathcal{Q}') + H(\mathcal{Q}|\mathcal{Q}')
$$

we can then conclude as before that $h(T, \mathcal{Q}) \leq h(T, \mathscr{P})$. It follows that if $\mathscr{P}$ is a generating partition, meaning that the partitions $T^k \mathscr{P}$ for $k \in \mathbb{Z}$ generate the σ-algebra modulo null sets, then

$$
h(T) = h(T, \mathscr{P}),
$$

which is known as the Kolmogorov–Sinai theorem. A more general version of this theorem, which follows by the same kind of perturbation argument, states that if $\mathscr{P}_1 \leq \mathscr{P}_2 \leq \ldots$ is a sequence of finite partitions whose images under G collectively generate the σ-algebra modulo null sets, then

$$
h(T) = \sup_{n \in \mathbb{N}} h(T, \mathscr{P}_n).
$$

The Kolmogorov–Sinai theorem applies in a prototypical way to a Bernoulli shift $T : Y^{\mathbb{Z}} \to Y^{\mathbb{Z}}$, where Y is a finite set (the *base* of the Bernoulli shift), $Y^{\mathbb{Z}}$ is equipped with some product probability measure $\nu^{\mathbb{Z}}$, and $(Ty)_n = y_{n-1}$ for all $y = (y_n) \in Y^{\mathbb{Z}}$, to show that the entropy $h(T)$ is equal to the entropy with respect to the canonical generating partition $\mathscr{P}$ consisting of the cylinder sets $\{(y_n) \in Y^{\mathbb{Z}} : y_0 = a\}$ for $a \in Y$. Since for every n we have $H(\mathscr{P} \vee T^{-1} \mathscr{P} \vee \cdots \vee T^{-n-1} \mathscr{P}) = nH(\mathscr{P})$ by the independence of the images of $\mathscr{P}$ under T, we obtain the simple formula

$$
h(T) = H(\mathscr{P}) = \sum_{a \in Y} -\nu(\{a\}) \log \nu(\{a\}).
$$

94   *Recent Advances in Operator Theory and Operator Algebras*

The sum on the right is referred to as the *base entropy* and is simply the Shannon entropy of the partition of Y into its singletons.

Now that we have seen how the Følner property underpins the basic theory of entropy for $\mathbb{Z}$-actions, we can repeat the above discussion for a p.m.p. action $G \curvearrowright (X, \mu)$ of a general countable amenable group by taking a Følner sequence $\{F_n\}$ for G and setting, for a finite partition $\mathscr{P}$,

$$h(\mathscr{P}) = \lim_{n \to \infty} \frac{1}{|F_n|} H\left(\bigvee_{s \in F_n} s^{-1} \mathscr{P} \right) \tag{3.2}$$

and then defining the *measure entropy* of the action by

$$h(X, G) = \sup_{\mathscr{P}} h(\mathscr{P}),$$

where $\mathscr{P}$ ranges over all finite partitions of X. We also write $h_\mu(X, G)$ if μ needs to be stressed. The Kolmorogov–Sinai theorem, which as before can be established using the easily verified inequality

$$h(\mathscr{Q}) \leq h(\mathscr{Q}') + H(\mathscr{Q} | \mathscr{Q}'), \tag{3.3}$$

asserts that if $\mathscr{P}$ is a generating partition, meaning that the partitions $s\mathscr{P}$ for $s \in G$ generate the σ-algebra modulo null sets, then $h(X, G) = h(\mathscr{P})$, and also more generally that $h(X, G) = \sup_{n \in \mathbb{N}} h(\mathscr{P}_n)$ whenever $\mathscr{P}_1 \leq \mathscr{P}_2 \leq \ldots$ are finite partitions whose images under G collectively generate the σ-algebra modulo null sets. For a Bernoulli action $G \curvearrowright (Y^G, \nu^G)$, defined by $(sy) = y_{s^{-1}t}$ for $s \in G$ and $y = (y_t) \in Y^G$, with Y and ν as before, we then take the canonical generating partition $\mathscr{P}$ consisting of the cylinder sets $\{(y_t) \in Y^G : y_e = a\}$ for $a \in Y$ to obtain the same formula

$$h(Y^G, G) = H(\mathscr{P}) = \sum_{a \in Y} -\nu(\{a\}) \log \nu(\{a\}).$$

This formula also holds for any atomic base probability space (Y, ν). For base probability spaces (Y, ν) which are not atomic, the entropy is infinity.

One can moreover show, using a strong form of subadditivity satisfied by Shannon entropy, that the limit in (3.2) does not depend on the choice of Følner sequence, and also that this limit is equal to

$$\inf_F \frac{1}{|F|} H\left(\bigvee_{s \in F} s^{-1} \mathscr{P} \right),$$

where F ranges over the nonempty finite subsets of G (see Section 9.3 of [36]). This latter expression is very useful in showing, for example, that if $G \overset{\alpha}{\curvearrowright} (X, \mu)$ is a p.m.p. action and $\mathscr{C}$ a G-invariant sub-σ-algebra of the given σ-algebra, then

$$h(\alpha) = h(\alpha|_{\mathscr{C}}) + h(\alpha | \mathscr{C})$$

(see Theorem 8.16 of [36]).

One of the most celebrated and influential achievements in ergodic theory is Ornstein's isomorphism theorem, which says that two Bernoulli shifts with the same (base) entropy are conjugate [44], so that (base) entropy is a complete invariant for Bernoulli shifts. This entropy classification was later generalized by Ornstein and Weiss to Bernoulli actions over arbitrary countably infinite amenable groups [45], and then by Bowen to the sofic context, as we will discuss in Section 3.5.

Despite the successes in extending much of abstract entropy theory for $\mathbb{Z}$ to the general amenable setting, in many domains of application, notably those involving smooth dynamics, the utility of dynamical entropy is essentially restricted to the case of single transformations. A smooth $\mathbb{Z}^2$-action, for example, always has zero entropy, since diffeomorphisms always have finite entropy and it is not hard to check that nonzero entropy for a $\mathbb{Z}^2$-action implies infinite entropy for the transformations corresponding to each nonzero group element. On the other hand, amenable entropy is tailor-made for algebraic actions, as we will see in Section 3.8.

3.4 Amenable topological entropy

Let X be a compact metrizable space. Given an open cover $\mathscr{U}$ of X, we denote by $N(\mathscr{U})$ the minimum cardinality of a subcover of $\mathscr{U}$. The following properties are then immediate, where $\mathscr{U}$ and $\mathscr{V}$ are finite open covers:

1. If $\mathscr{V}$ refines $\mathscr{U}$ then $N(\mathscr{V}) \geq N(\mathscr{U})$.

2. If $T : X \to X$ is a homeomorphism then $N(T\mathscr{U}) = N(\mathscr{U})$.

3. $N(\mathscr{U} \vee \mathscr{V}) \leq N(\mathscr{U})N(\mathscr{V})$.

For a homeomorphism $T : X \to X$ and a finite open cover $\mathscr{U}$ of X we set

$$h_{\text{top}}(T, \mathscr{U}) = \lim_{n \to \infty} \frac{1}{n} \log N(\mathscr{U} \vee T^{-1}\mathscr{U} \vee \cdots \vee T^{-n-1}\mathscr{U}).$$

The fact that the limit exists can be deduced from the submultiplicativity in (3). The *topological entropy* of T is then defined by taking a supremum over all finite open covers:

$$h_{\text{top}}(T) = \sup_{\mathscr{U}} h_{\text{top}}(T, \mathscr{U}).$$

This definition was introduced by Adler, Konheim, and McAndrew in the 1960s [1]. The analogue of the Kolomogorov–Sinai in this setting says that if $\mathscr{U}$ is generating, i.e., the sets $T^n\mathscr{U}$ for $n \in \mathbb{Z}$ generate the topology, then

$h_{\text{top}}(T) = h_{\text{top}}(\mathscr{U})$. For a shift $T : \{1, \ldots, d\}^{\mathbb{Z}} \to \{1, \ldots, d\}^{\mathbb{Z}}$ defined by $(Ty)_n = y_{n-1}$ this implies, by taking $\mathscr{U}$ to be the clopen partition of X consisting of the cylinder sets $\{(y_n) \in \{1, \ldots, d\}^{\mathbb{Z}} : y_0 = k\}$ for $k = 1, \ldots, d$, that

$$h_{\text{top}}(T) = h_{\text{top}}(T, \mathscr{U}) = \log d.$$

We now generalize as before by taking a continuous action $G \curvearrowright X$ of a countable amenable group with Følner sequence $\{F_n\}$ and, for a finite open cover $\mathscr{U}$ of X, setting

$$h_{\text{top}}(\mathscr{U}) = \lim_{n \to \infty} \frac{1}{n} \log N \left(\bigvee_{s \in F_n} s^{-1} \mathscr{U} \right).$$

That the limit exists and is independent of the choice of Følner sequence can be shown using the submultiplicativity property (3) (see Section 9.9 of [36]). The *topological entropy* of the action is then defined by

$$h_{\text{top}}(X, G) = \sup_{\mathscr{U}} h_{\text{top}}(\mathscr{U}).$$

As before, for the shift action $G \curvearrowright \{1, \ldots, q\}^G$ defined by $(sy)_t = y_{s^{-1}t}$ for $s \in G$ and $y \in \{1, \ldots, q\}^G$ we have, using the canonical generating clopen partition,

$$h_{\text{top}}(X^G, G) = \log q.$$

As an application of topological entropy which we will revisit later in the sofic context, we will give a proof of Gottschalk's surjunctivity conjecture in the case of amenable groups. The customary definition of finiteness for a set A requires that there exist a bijection from A to a set of the form $\{1, \ldots, q\}$ for some natural number q. Under the axiom of choice this is equivalent to Dedekind finiteness, which requires that every injective map from A to itself be surjective. In [24] Gottschalk wondered whether every countable group G has the property that for every $q \in \mathbb{N}$ the compact space $\{1, \ldots, q\}^G$ satisfies a G-equivariant version of Dedekind finiteness with respect to the (left) shift action $(sy)_t = y_{s^{-1}t}$, meaning that every injective G-equivariant continuous map from $\{1, \ldots, q\}^G$ to itself is automatically surjective. While this problem is still open, the property in question, called *surjunctivity*, was shown to hold for sofic groups by Gromov, and in fact was Gromov's motivation for introducing the idea of soficity [25]. As Gromov points out, surjunctivity can be verified for amenable groups using topological entropy, as we now explain.

Let G be a countable amenable group, $q \in \mathbb{N}$, and $G \curvearrowright \{1, \ldots, q\}^G$ the shift action. Let $\varphi : \{1, \ldots, q\}^G \to \{1, \ldots, q\}^G$ be an injective G-equivariant map, and write Y for the image of φ. Then Y is a closed G-invariant subset of $\{1, \ldots, q\}^G$, and φ establishes a conjugacy between the full shift $G \curvearrowright \{1, \ldots, q\}^G$ and the subshift action $G \curvearrowright Y$, so that this subshift has the same entropy as the full shift, namely $\log q$. Assuming now that Y is a proper subset of $\{1, \ldots, q\}^G$, the argument will be finished, via contradiction, upon showing that the entropy of $G \curvearrowright Y$ is strictly less than $\log q$, as we now proceed to do. For the purpose of expressing entropy we fix a Følner sequence $\{F_n\}$ for G.

Since Y is a proper closed G-invariant subset, we can find a finite set $K \subseteq G$ and a map $f : K \to \{1, \ldots, q\}$ such that for every $s \in G$ the map $st \mapsto f(t)$ on sK does not appear as the restriction of any element of Y. By the Følner condition, as expressed in (3.1), if n is large enough then the set $A = \{s \in F_n : Ks \subseteq F_n\}$ satisfies $|A| \geq |F_n|/2$. Take a maximal set $A_0 \subseteq A$ such that the sets Ks for $s \in A_0$ are pairwise disjoint. Note that if $s, t \in A$ satisfy $Ks \cap Kt \neq \emptyset$ then $s \in K^{-1}Kt$, which implies that $|A_0| \geq |A|/|K^{-1}K| \geq |F_n|/(2|K|^2)$. Observe also that the set of all maps $F_n \to \{1, \ldots, q\}$ which are restrictions of elements of Y to F_n has cardinality at most

$$q^{|F_n| - |K||A_0|}(q^{|K|} - 1)^{|A_0|}.$$

It follows, writing $\mathscr{U}$ for the canonical generating clopen partition consisting of the cylinder sets $\{(y_n) \in \{1, \ldots, q\}^{\mathbb{Z}} : y_0 = k\}$ for $k = 1, \ldots, q$ and setting

$$\eta = |K| \log q - \log(q^{|K|} - 1) > 0,$$

that

$$\frac{1}{|F_n|} \log N \left(\bigvee_{s \in F_n} s^{-1} \mathscr{U} \right) \leq \left(1 - \frac{|K||A_0|}{|F_n|} \right) \log q + \frac{|A_0|}{|F_n|} (|K| \log q - \eta)$$

$$\leq \log q - \frac{\eta}{2|K|^2}.$$

This upper bound is strictly less than $\log q$ and does not depend on n, so that the entropy of $G \curvearrowright Y$ is strictly less than $\log q$, giving the desired contradiction that finishes the argument.

We next explain an alternative "dual" formulation of topological entropy in terms of separated and spanning sets with respect to a metric, or more generally with respect to a dynamically generating pseudometric. This viewpoint, which was originally introduced for $\mathbb{Z}$-actions by Bowen [9] and Dinaburg [14], is often more convenient when carrying out entropy computations or estimates on spaces of nonzero covering dimension, for which the problem cannot be immediately combinatorialized as in the subshift setting above. One can interpret the meaning of "duality" here in the C*-algebraic terms of the Gelfand transform: we are trading finite open covers of X, which in the simplest case of clopen partitions correspond to finite-dimensional subalgebras of $C(X)$, for finite sets of points in X, which correspond to finite-dimensional quotients of $C(X)$.

Let $G \curvearrowright X$ be a continuous action of a countable amenable group on a compact metrizable space. Let ρ be a continuous pseudometric on X.

Definition 3.4.1. *Let F be a nonempty finite subset of G. On X we define the pseudometric*

$$\rho_F(x, y) = \max_{s \in F} \rho(sx, sy),$$

and say that a set $D \subseteq X$ is (ρ, F, ε)-separated if $\rho_F(x, y) \geq \varepsilon$ for all distinct $x, y \in D$, and (ρ, F, ε)-spanning if for every $x \in X$ there is a $y \in D$ such that $\rho_F(x, y) < \varepsilon$. Write $\mathrm{sep}(\rho, F, \varepsilon)$ for the maximum cardinality of a (ρ, F, ε)-separated subset of X and $\mathrm{spn}(\rho, F, \varepsilon)$ for the minimum cardinality of a (ρ, F, ε)-spanning subset of X, and set

$$h_{\mathrm{spn}}(\rho) = \sup_{\varepsilon > 0} \limsup_{n \to \infty} \frac{1}{|F_n|} \log \mathrm{spn}(\rho, F_n, \varepsilon),$$

$$h_{\mathrm{sep}}(\rho) = \sup_{\varepsilon > 0} \limsup_{n \to \infty} \frac{1}{|F_n|} \log \mathrm{sep}(\rho, F_n, \varepsilon).$$

We say that the pseudometric ρ is *dynamically generating* if for all distinct $x, y \in X$ there is an $s \in G$ such that $\rho(sx, sy) > 0$.

Theorem 3.4.2. *Suppose that ρ is dynamically generating. Then*

$$h_{\mathrm{top}}(X, G) = h_{\mathrm{spn}}(\rho) = h_{\mathrm{sep}}(\rho).$$

Note that while there is no canonical compatible metric on the product space $\{1, \ldots, d\}^G$, there is a more or less canonical generating pseudometric ρ for the shift action $G \curvearrowright \{1, \ldots, d\}^G$, namely the one whose value is 0 or 1 depending on whether the coordinates of the two points in question at e agree or disagree. Using this ρ one can compute the entropy just as before to be $\log d$.

Finally we mention the variational principle connecting measure and topological entropy [13, 14, 22, 23, 43].

Theorem 3.4.3. *Let $G \curvearrowright X$ be a continuous action of a countable amenable group on a compact metrizable space. Then*

$$h_{\mathrm{top}}(X, G) = \sup_{\mu} h_{\mu}(X, G),$$

where μ ranges over all G-invariant Borel probability measures on X.

3.5 Sofic measure entropy

In the definition of Shannon entropy in Section 3.3, we used an internal information-theoretic heuristic that only made reference to subsets of the space and their measures. The Følner characterization amenability for a group is a similarly internal property, and the asymptotic invariance that it expresses is exactly what was needed to define a viable invariant, namely amenable measure entropy, through the asymptotic averaging of Shannon entropies over the dynamics.

An introduction to sofic entropy 99

There is a natural bijective correspondence between finite partitions of a probability space (X, μ) and finite sub-σ-algebras (which are the same thing as finite (Boolean) subalgebras) of the σ-algebra $\mathscr{B}_X$ of (X, μ) that associates to such a partition the subalgebra it generates, and so from the algebraic viewpoint one can view amenable measure entropy as taking a limit of averages that are computed directly from the numerical characteristics of finite subalgebras of $\mathscr{B}_X$ (i.e., the measures of their "atoms"). One could try to externalize this picture by mapping $\mathscr{B}_X$ to the algebra $\mathbb{P}_d$ of all subsets of $\{1, \ldots, d\}$ for some $d \in \mathbb{N}$. In order to generate an entropy-type quantity out of this, if it were indeed possible to do so, there has to be some locality involved, which in the Shannon/Kolmogorov–Sinai setting is expressed by the finiteness of the subalgebra. We could thus fix a subalgebra of $\mathscr{B}_X$ and map it homomorphically in a measure-preserving or almost measure-preserving way to $\mathbb{P}_d$. The one point of uncertainty opened up by this external viewpoint is that there are many such homomorphisms to $\mathbb{P}_d$, none of which is more canonical than others. It is precisely this uncertainty that we will exploit to define an alternative notion of entropy, an idea that originates in the work of Boltzmann in statistical mechanics. At the spatial level, this means that we will count the number of ways a finite ordered partition of X can be modeled by a finite ordered partition of $\{1, \ldots, d\}$ so that the measures of partition members that correspond to each other under the orderings have approximately the same measure.

The basic problem is really then to determine, given $\lambda_1, \ldots, \lambda_n > 0$ with $\sum_{i=1}^{n} \lambda_i = 1$, how many ordered n-element "ordered" partitions $\mathcal{Q} = \{B_1, \ldots, B_n\}$ of $\{1, \ldots, d\}$ have distribution roughly $(\lambda_1, \ldots, \lambda_n)$, i.e., $|B_i|/d \approx_\varepsilon \lambda_i$ for each i and some prescribed $\varepsilon > 0$. If the numbers $\lambda_1 d, \ldots, \lambda_n d$ happen to be integers, then the number of ordered partitions of $\{1, \ldots, d\}$ with distribution exactly $(\lambda_1, \ldots, \lambda_n)$ is equal to

$$\frac{d!}{(\lambda_1 d)! \cdots (\lambda_n d)!}, \tag{3.1}$$

which, after taking a logarithm and dividing by d, we refer to as the *Boltzmann entropy*. In statistical mechanics (3.1) represents the number of ways that n different states can be distributed among d particles subject to the constraint that there be a fixed number $\lambda_i d$ particles in the ith state for every i. That this is (however paradoxically) a physically meaningful quantity is explained by the fact that, while we are able to observe the distribution macroscopically (i.e., we can determine the numbers λ_i through observation), the particles themselves are physically indistinguishable, so that we are unable to observe any individual assignment of states to particles ("microstates"). Using Stirling's formula $m! \sim \sqrt{2\pi m}(m/e)^m$, we see that the logarithm of (3.1), when normalized by dividing by d, tends, as $d \to \infty$, to

$$\sum_{i=1}^{n} -\lambda_i \log \lambda_i,$$

100   Recent Advances in Operator Theory and Operator Algebras

which is precisely the Shannon entropy of any partition with distribution $(\lambda_1, \ldots, \lambda_n)$. By using some more combinatorics and Stirling's formula again, one can moreover see that the number of ordered partitions of $\{1, \ldots, d\}$ with distribution approximately equal to $(\lambda_1, \ldots, \lambda_n)$ in the above sense, as expressed in terms of ε, is approximately equal to $\sum_{i=1}^{n} -\lambda_i \log \lambda_i$ to within some $\eta > 0$, which tends to 0 as $\varepsilon \to 0$. Therefore the limit of (approximate) Boltzmann entropies is (approximately) equal to the Shannon entropy.

While the Boltzmann entropy thus merely reproduces the Shannon entropy if we view it, in asymptotic form, as a numerical invariant of a fixed finite partition of our space X, when coupled with a group action it opens up a very broad realm of application that takes us well beyond the scope of amenability. In order to produce a dynamical invariant for a p.m.p. action $G \curvearrowright (X, \mu)$, we use the finite set $\{1, \ldots, d\}$ to model not only a given finite partition $\mathscr{P}$ of X but also its evolution under the action. We would then need an action of G on the set $\{1, \ldots, d\}$ with respect to which this dynamical modeling can be done. In fact, it turns out to be enough to have "almost" actions of G on $\{1, \ldots, d\}$ which are "almost" free in a sense that is made precise in the following definition of a sofic group. Recall that $\mathrm{Sym}(d)$ denotes the permutation group of the set $\{1, \ldots, d\}$.

Definition 3.5.1. *The group G is said to be sofic if there are a sequence $\{d_i\}$ of natural numbers and a sequence of maps $\sigma_i : G \to \mathrm{Sym}(d_i)$ which are asymptotically multiplicative and asymptotically free in the sense that*

1. $\lim_{i \to \infty} \left| \{v \in \{1, \ldots, d_i\} : \sigma_{i,st}(v) = \sigma_{i,s}\sigma_{i,t}(v)\} \right| / d_i = 1$ *for all $s, t \in G$, and*

2. $\lim_{i \to \infty} \left| \{v \in \{1, \ldots, d_i\} : \sigma_{i,s}(v) \neq \sigma_{i,t}(v)\} \right| / d_i = 1$ *for all distinct $s, t \in G$,*

where $\sigma_{i,s}$ is the image of the group element s under σ_i. Such a sequence $\{\sigma_i\}$ with $d_i \to \infty$ is called a sofic approximation sequence for G (the requirement that $d_i \to \infty$ is automatic when G is infinite and helps us to avoid pathologies in entropy theory when G is finite).

We will also refer to a single map $\sigma : G \to \mathrm{Sym}(d)$ as a *sofic approximation*, and also speak of such maps as being good or good enough sofic approximations in the sense that, for whatever the purpose at hand, the averages in (1) and (2) above are close enough to the limiting value 1 for s and t ranging in a suitable finite subset of G (with $s \neq t$ in (2)).

Example 3.5.2. *(1). Recall that a group G is residually finite if for every finite set $F \subseteq G$ there is a finite group H and a homomorphism $\varphi : G \to H$ such that the restriction of φ to F is injective. Such a group is sofic, for we can take a sequence $\{K_i\}$ of finite-index subgroups of G such that for every $s \in G \setminus \{e\}$ we eventually have $sK_i \neq K_i$ in the quotient G/K_i, so that the maps σ_i from G to the permutation group $\mathrm{Sym}(G/K_i)$ given by $\sigma_{i,s}(tK_i) = stK_i$ form a sofic* *approximation sequence, with the averages in (1) of Definition 3.5.1 always being equal to 1 (i.e., each σ_i is a genuine group action) and the averages in (2) for fixed s and t being exactly equal to 1 for all large enough i.*

The group $\mathbb{Z}$ is residually finite, as witnessed by the finite quotients $\mathbb{Z}/m\mathbb{Z}$ for $m \in \mathbb{N}$. Similarly, $\mathbb{Z}^r$ for $r \in \mathbb{N}$ is residually finite. Another example is the special linear group $\mathrm{SL}(n, \mathbb{Z})$ (where $n \in \mathbb{N}$) of invertible $n \times n$ integer matrices having determinant 1, as we can consider for every $k \in \mathbb{N}$ the finite-index normal subgroup of all elements whose nondiagonal entries belong to $k\mathbb{Z}$ and whose diagonal entries belong to $1 + k\mathbb{Z}$. The free group F_r on r generators, where $r \in \mathbb{N}$, is also residually finite. There are several different ways of verifying this, one of which is to show that F_r embeds into $\mathrm{SL}(2, \mathbb{Z})$ (see Section II.B of [11]).

(2). Amenable groups are sofic. Indeed if $\{F_i\}$ is a Følner sequence for an amenable group G then for each i we define a map $\sigma_i : G \to \mathrm{Sym}(F_i)$ by setting $\sigma_{i,s}t = st$ for all $s \in G$ and $t \in F$ satisfying $st \in F_i$ and then completing the definition of each $\sigma_{i,s}$ in any way that produces a bijection from F_i to itself. Then the Følner condition guarantees that $\{\sigma_i\}$ is a sofic approximation sequence. Amenable groups can be moreover characterized by the fact that any two sofic approximations on a finite set are approximately conjugate to each other by a permutation [16]. This is a kind of uniqueness result that is very much analogous to the uniqueness theorems in the classification theory of simple separable nuclear (i.e., amenable) C^*-algebras, which give approximate uniqueness for various maps up to approximate innerness.

(3). One can simultaneously generalize the above two classes by talking about residually amenable groups, which one obtains by replacing the finite groups in (1) with arbitrary amenable groups. For examples of sofic groups which are not residually amenable see [16].

It remains unknown whether nonsofic groups exist.

Unless we state otherwise, from this point onward the group G is sofic and $\Sigma = \{\sigma_i : G \to \mathrm{Sym}(d_i)\}$ is a fixed sofic approximation sequence for G. We also assume a p.m.p. action $G \curvearrowright (X, \mu)$ to be given.

Notation. For a finite partition $\mathscr{P}$ of X and a finite set $F \subseteq G$, we write $\mathscr{P}_F$ for the join $\bigvee_{s \in F} s\mathscr{P}$. For a finite partition of X we write $\mathrm{alg}(\mathscr{C})$ for the algebra generated by $\mathscr{C}$, which as a set consists of all unions of members of $\mathscr{C}$. As before, we write $\mathbb{P}_d$ for the algebra of all subsets of $\{1, \ldots, d\}$.

Definition 3.5.3. *Let $\mathscr{C}$ be a finite partition of X, F a finite subset of G containing e, and $\delta > 0$. Let σ be a map from G to $\mathrm{Sym}(d)$ for some $d \in \mathbb{N}$. We write $\mathrm{Hom}_\mu(\mathscr{C}, F, \delta, \sigma)$ for the set of all homomorphisms $\varphi : \mathrm{alg}(\mathscr{C}_F) \to \mathbb{P}_d$ such that*

1. *$\sum_{A \in \mathscr{C}} |\sigma_s \varphi(A) \Delta \varphi(sA)|/d < \delta$ for all $s \in F$, and*

2. *$\sum_{A \in \mathscr{C}_F} |\varphi(A)/d - \mu(A)| < \delta$.*

102   *Recent Advances in Operator Theory and Operator Algebras*

Moreover, for a finite partition $\mathscr{P}$ which $\mathscr{C}$ refines we write $|\operatorname{Hom}_\mu(\mathscr{C}, F, \delta, \sigma)|_\mathscr{P}$ for the cardinality of the set of restrictions of homomorphisms in $\operatorname{Hom}_\mu(\mathscr{C}, F, \delta, \sigma)$ to $\mathscr{P}$ (note that $\mathscr{P} \subseteq \operatorname{alg}(\mathscr{C}_F)$ because $e \in F$).

The homomorphisms in $\operatorname{Hom}_\mu(\mathscr{C}, F, \delta, \sigma)$ are determined by their restriction to $\mathscr{C}_F$, which maps members of one partition to members of another. In what follows one could alternatively adopt this viewpoint, in which case the restriction of a homomorphism in $\operatorname{Hom}_\mu(\mathscr{C}, F, \delta, \sigma)$ to a partition $\mathscr{P} \leq \mathscr{C}$ would be expressed as the corresponding "coarsening" of the maps between the underlying partitions (see Section 7 of [20] or Section 1.8 of [19]). The algebraic perspective makes it clear that we are dealing with "approximate" or "local" morphisms of group actions, from the action of G on X to an (approximate) action of G on a finite set. It is because these morphisms are localized to finite subalgebras that they admit a combinatorial description in terms of partitions. For another approach where the localization is expressed in a linear/operator-algebraic way, see [33].

Definition 3.5.4. *For a finite partition $\mathscr{P}$ of X we set*

$$h_\Sigma(\mathscr{P}) = \inf_{\mathscr{C} \geq \mathscr{P}} \inf_F \inf_{\delta > 0} \limsup_{i \to \infty} \frac{1}{d_i} \log |\operatorname{Hom}_\mu(\mathscr{C}, F, \delta, \sigma_i)|_\mathscr{P},$$

where $\mathscr{C}$ ranges over all finite partitions of X refining $\mathscr{P}$ and F ranges over all finite sets $F \subseteq G$ containing e. When $\mathscr{C}$, F, and δ are fixed we write $h_\Sigma(\mathscr{P}, \mathscr{C}, F, \delta)$ for the limit supremum above, and when $\mathscr{C}$ is fixed we write $h_\Sigma(\mathscr{P}, \mathscr{C})$ for the infimum over F and δ. We then define the entropy of the action $G \curvearrowright (X, \mu)$ with respect to Σ as

$$h_\Sigma(X, G) = \sup_\mathscr{P} h_\Sigma(\mathscr{P}),$$

where $\mathscr{P}$ ranges over all finite partitions of X. If the measure needs to be emphasized then we write $h_{\Sigma,\mu}(\mathscr{P})$, $h_{\Sigma,\mu}(X, G)$, etc.

The idea of the above definition is that $\mathscr{C}$, F, and δ control how good the dynamical modeling is, while $\mathscr{P}$ describes the window or scale of observation. One can think of the role of $\mathscr{P}$ as being analogous to that of the tolerance ε in the pseudometric formulation of topological entropy in Section 3.4.

It can certainly happen that $\operatorname{Hom}_\mu(\mathscr{C}, F, \delta, \sigma_i)$ is empty, in which case we interpret $\log 0$ as $-\infty$ for the purpose of taking infima and suprema. Thus the range of possible values for $h_\Sigma(X, G)$ is $\{-\infty\} \cup [0, \infty]$, which differs from the amenable case in including $-\infty$.

It is also worth noting that, unlike in the amenable case, the entropy $h_\Sigma(\mathscr{P})$ with respect to the partition $\mathscr{P}$ does not merely depend on the G-invariant σ-algebra generated by $\mathscr{P}$, for $\mathscr{C}$ ranges over *all* finite partitions refining $\mathscr{P}$ in the definition of $h_\Sigma(\mathscr{P})$. Indeed it is necessary for this to be the case as it will turn out that sofic entropy can increase undertaking factors. This nonmonotone behavior is a major point of difference with amenable entropy.

An introduction to sofic entropy 103

Example 3.5.5. *As an exercise one can compute the sofic entropy of the trivial action $G \curvearrowright (X, \mu)$ (i.e., $sx = x$ for all $s \in G$ and $x \in X$) to be zero. For amenable entropy this is immediate from the definition, but the sofic case requires a nontrivial combinatorial argument. The trivial action is a trivial example of a p.m.p. action $G \curvearrowright (X, \mu)$ which is compact, meaning that if we consider the unitary representation $\kappa : G \to L^2(X)$ given by $\pi(s)\xi(x) = \xi(s^{-1}x)$ (the Koopman representation) then the closure $\overline{\pi(G)\xi}$ is compact in the L^2 norm. One can show more generally that the sofic entropy of compact p.m.p. action $G \curvearrowright (X, \mu)$ is always 0 or $-\infty$ (Theorem 10.18 of [36]). It is also a good exercise in the use of conditional Shannon entropy to compute the amenable entropy of such an action to be zero when G is amenable.*

The following is the counterpart of (3.3) from Section 3.3, and is useful for example in computing the entropy of Bernoulli actions.

Proposition 3.5.6. *For finite partitions $\mathscr{P}$ and $\mathscr{Q}$ of X one has*

$$h_\Sigma(\mathscr{P}) \leq h_\Sigma(\mathscr{Q}) + H(\mathscr{P}|\mathscr{Q}),$$

and in particular, taking $\mathscr{Q}$ to be the trivial partition, $h_\Sigma(\mathscr{P}) \leq H(\mathscr{P})$.

Another ingredient we need for computing the entropy of Bernoulli actions, like in the amenable setting, is the following generator theorem, which is the analogue of the Kolmogorov–Sinai theorem. We say that a subalgebra $\mathscr{S}$ of the σ-algebra $\mathscr{B}_X$ of (X, μ) is *generating* if the smallest G-invariant sub-σ-algebra of $\mathscr{B}_X$ containing $\mathscr{S}$ is, modulo null sets, $\mathscr{B}_X$ itself.

Theorem 3.5.7. *Let $\mathscr{S}$ be a generating subalgebra of $\mathscr{B}_X$. Then*

$$h_\Sigma(X, G) = \sup_{\mathscr{P}} h_\Sigma(\mathscr{P}),$$

where $\mathscr{P}$ ranges over all finite partitions contained in $\mathscr{S}$.

The proof of this is quite a bit more involved than the corresponding amenable version, and does not (and cannot) contain the kind of unidirectional conclusion $h(\mathscr{Q}) \leq h(\mathscr{P})$ that one can derive in the amenable case assuming that $\mathscr{Q}$ is contained, modulo null sets, in the G-invariant σ-algebra generated by $\mathscr{P}$. The argument requires a kind of back-and-forth that in its most basic form shows that $h(\mathscr{Q}) = h(\mathscr{P})$ whenever $\mathscr{P}$ and $\mathscr{Q}$ are *both* generating partitions. This invariance with respect to generating partitions was originally established by Bowen, who used it to give the first definition of sofic entropy. Since not all p.m.p. actions admit generating partitions (for example, the trivial action on an atomless space), Definition 3.5.4 is needed in general, and we know by Theorem 3.5.7 that this reduces to Bowen's definition in the generating case.

With Proposition 3.5.6 and Theorem 3.5.7 at hand, one can compute the entropy of a Bernoulli action to be the Shannon entropy of the base [5], as one expects.

104 Recent Advances in Operator Theory and Operator Algebras

Theorem 3.5.8. *Let $n \in \mathbb{N}$ and let ν be a probability measure on a finite set Y. Then for the Bernoulli action $G \curvearrowright (Y^G, \nu^G)$ one has*

$$h_\Sigma(Y^G, G) = \sum_{a \in Y} -\nu(\{a\}) \log \nu(\{a\}).$$

Unlike in the amenable case, and not surprisingly in view of our discussion of the trivial action in Example 3.5.5, the computation in Theorem 3.5.8 requires additional combinatorial work beyond an appeal to Proposition 3.5.6 and Theorem 3.5.7. The proof proceeds by a probabilistic second moment argument that uses the law of large numbers and Chebyshev's inequality.

One can go further to produce an entropy classification of Bernoulli actions as in the amenable setting. Stepin observed using a coinduction argument that Ornstein's isomorphism theorem holds for a given countable group whenever it holds for one of its subgroups. Using a similar kind of coinduction argument, Bowen proved that Ornstein's isomorphism theorem holds for every countably infinite group if one omits the Bernoulli actions with two-atom bases, which together with Theorem 3.5.8 yields a nearly complete entropy classification of Bernoulli actions of countably infinite sofic groups [7]. Bowen also gave in [7] a complete entropy classification of Bernoulli actions for a large class of sofic groups.

It was shown in [31] that Bernoulli actions have *complete positive entropy*, meaning that every nontrivial factor has nonzero entropy, a fact which was well known in the amenable case.

When the acting group is amenable, sofic measure entropy coincides with amenable measure entropy for every choice of sofic approximation sequence [8, 34]. The proof given in [34] relies on the fact that every good-enough sofic approximation for an amenable group can be approximately tiled by partial orbits over Følner sets with prescribed approximate invariance. This is an "externalized" version of the Ornstein–Weiss quasitiling theorem for subsets of the group itself [45] and is established by the same argument.

Theorem 3.5.9. *Suppose that G is amenable, and let $G \curvearrowright (X, \mu)$ be a p.m.p. action. Let Σ be a sofic approximation sequence for G. Then*

$$h_{\Sigma,\mu}(X, G) = h_\mu(X, G).$$

3.6 Sofic topological entropy

Although one can define sofic topological entropy using open covers as we originally did in the amenable case [53], we will instead concentrate here on the pseudometric approach.

Throughout G is a countable sofic group, $\Sigma = \{\sigma_i : G \to \operatorname{Sym}(d_i)\}$ is a sofic approximation sequence, $G \curvearrowright X$ is a continuous action on a compact metrizable space, and ρ is a continuous pseudometric on X.

Given $d \in \mathbb{N}$, we define on the set of all maps $\{1, \dots, d\} \to X$ the pseudometrics

$$\rho_2(\varphi, \psi) = \left(\frac{1}{d} \sum_{v=1}^{d} (\rho(\varphi(v), \psi(v)))^2 \right)^{1/2},$$

$$\rho_\infty(\varphi, \psi) = \max_{v=1,\dots,d} \rho(\varphi(v), \psi(v)).$$

Definition 3.6.1. *Let F be a finite subset of G and $\delta > 0$, and let $\sigma : G \to \operatorname{Sym}(d)$ for some $d \in \mathbb{N}$. Define $\operatorname{Map}(\rho, F, \delta, \sigma)$ to be the set of all maps $\varphi : \{1, \dots, d\} \to X$ satisfying the approximate equivariance condition $\rho_2(\varphi\sigma_s, \alpha_s\varphi) \leq \delta$ for all $s \in F$, where α_s is the transformation $x \mapsto sx$ of X.*

We write $N_\varepsilon(Y, \rho)$ for the maximum cardinality of a subset of Y which is (ρ, ε)-separated, i.e., $\rho(y, z) \geq \varepsilon$ for all distinct $y, z \in Y$.

Definition 3.6.2. *We define*

$$h_\Sigma(\rho) = \sup_{\varepsilon > 0} \inf_F \inf_{\delta > 0} \limsup_{i \to \infty} \frac{1}{d_i} \log N_\varepsilon(\operatorname{Map}(\rho, F, \delta, \sigma_i), \rho_\infty),$$

where F ranges over the finite subsets of G.

Proposition 3.6.3. *If ρ and ρ' are two dynamically generating continuous pseudometrics on X then $h_\Sigma(\rho) = h_\Sigma(\rho')$.*

One can also show that the value of $h_\Sigma(\rho)$ remains unchanged if in Definition 3.6.2 one substitutes ρ_2 for ρ_∞. The pseudometric ρ_∞ often turns out to be more convenient in applications.

Definition 3.6.4. *The topological entropy $h_\Sigma(X, G)$ of the action $G \curvearrowright X$ with respect to Σ is defined as the common value of $h_\Sigma(\rho)$ over all dynamically generating continuous pseudometrics ρ on X.*

The following is the variational principle in the sofic context [33].

Theorem 3.6.5. *Let $G \curvearrowright X$ be a continuous action of a countable sofic group on a compact metrizable space. Then*

$$h_\Sigma(X, G) = \sup_\mu h_{\Sigma,\mu}(X, G),$$

where μ ranges over all G-invariant Borel probability measures on X. In particular, if $h_\Sigma(X, G) \neq -\infty$ then the set of such μ is nonempty.

106   *Recent Advances in Operator Theory and Operator Algebras*

When the acting group is amenable, the sofic and amenable topological entropies coincide, just as in the measure case. The argument again makes use of quasitilings, although in the topological setting one does not need to negotiate the affine behavior of entropy with respect to decompositions, with the consequence that the proof is much simpler (while nevertheless being quite technical) [34].

Theorem 3.6.6. *Suppose that G is amenable, and let $G \curvearrowright X$ be a continuous action on a compact metrizable space. Let Σ be a sofic approximation sequence for G. Then*

$$h_\Sigma(X, G) = h_{\mathrm{top}}(X, G).$$

We finish this section by giving an entropy proof of the following result of Gromov [25], which resolves Gottschalk's surjunctivity conjecture for countable sofic groups. Recall that G is said to be *surjunctive* if, for every $q \in \mathbb{N}$, every injective G-equivariant continuous map from $\{1, \ldots, q\}^G$ to itself is surjective (see Section 3.4 for more discussion).

Theorem 3.6.7. *Every countable sofic group is surjunctive.*

Proof. Let G be a countable sofic group and let $G \curvearrowright \{1, \ldots, q\}^G$ be the usual (left) shift action. Let Y be a nonempty closed G-invariant proper subset of $\{1, \ldots, q\}^G$. Let Σ be a sofic approximation sequence for G. As in the argument for the amenable case in Section 3.4, it suffices to show that the entropy $h_\Sigma(Y, G)$ of the action $G \curvearrowright Y$ is strictly less than $\log q$. We will do this using the dynamically generating pseudometric ρ on Y which takes the value 0 or 1 depending on whether the coordinates of the two points in question agree or disagree at e.

Since Y is a proper closed G-invariant subset, there exist a finite set $K \subseteq G$ and a map $f : K \to \{1, \ldots, q\}$ which is not equal to the restriction of any element of Y. Let $\delta > 0$ be such that $\delta < 1/(2|K| + 4)$. Let $\sigma : G \to \mathrm{Sym}(d)$ and write V for the set of all $v \in \{1, \ldots, d\}$ such that the map from K to $\{1, \ldots, d\}$ given by $v \mapsto \sigma_{s^{-1}}(v)$ fails to be injective. We assume that σ is a sufficiently good sofic approximation to ensure that $|V| < \delta d$.

Let $0 < \varepsilon < 1$. Take a $(\rho_\infty, \varepsilon)$-separated set $\mathscr{M} \subseteq \mathrm{Map}(\rho, K^{-1}, \delta, \sigma)$ of maximum cardinality. We write $\mathscr{W}$ for the collection of all sets $W \subseteq \{1, \ldots, d\}$ containing V whose cardinality is exactly $\lceil (|K|+1)\delta d \rceil$. Let $W \in \mathscr{W}$ and define $\mathscr{M}_W$ to be the set of all $\varphi \in \mathscr{M}$ satisfying the local equivariance condition $\varphi(\sigma_{s^{-1}}(v))_e = (s^{-1}\varphi(v))_e$ for $v \in \{1, \ldots, d\} \setminus W$ and $s \in K$.

Observe that if we take a maximal set $U \subseteq \{1, \ldots, d\}$ with the property that the sets $\sigma(K^{-1})v$ for $v \in U$ are pairwise disjoint, then assuming d is sufficiently large we have

$$|U| \geq \frac{|\{1, \ldots, d\} \setminus W|}{|\sigma(K^{-1})^{-1}\sigma(K^{-1})|} \geq \frac{(1 - (|K| + 2)\delta)d}{|K|^2} \geq \frac{d}{2|K|^2}. \tag{3.1}$$

We next code the elements of $\mathscr{M}$ in an injective way by assigning to each

$\varphi \in \mathscr{M}$ the element ω_φ of $\{1, \ldots, k\}^d$ given by $\omega_\varphi(v) = \varphi(v)_e$. For every $\varphi \in \mathscr{M}_W$ and $s \in F$ we have

$$\omega_\varphi(\sigma_{s^{-1}}(v)) = (s^{-1}\varphi(v))_e = \varphi(v)_s,$$

which means that the map $s \mapsto \omega_\varphi(\sigma_{s^{-1}}(v))$ from K to $\{1, \ldots, d\}$ cannot be equal to f. It follows that for every $v \in U$ the number of different restrictions of the ω_φ to $\sigma(K^{-1})v$ is at most $q^{|K|} - 1$, whence

$$|\mathscr{M}_W| \leq q^{d-|K||U|}(q^{|K|} - 1)^{|U|} = (q^{|K|})^{d/|K|-|U|}(q^{|K|} - 1)^{|U|}. \tag{3.2}$$

In view of (3.1) the expression on the right is at most $q^{(1-\beta)d}$ for some $\beta > 0$ not depending on W, δ, d, or σ.

The total number of sets $W \subseteq \{1, \ldots, d\}$ with cardinality exactly $\lceil(|K| + 1)\delta d\rceil$ has cardinality $\binom{d}{\lceil(|K|+1)\delta d\rceil}$, and by Stirling's formula this binomial coefficient is bounded above, for all large-enough d, by $e^{\eta d}$ for some $\eta > 0$ not depending on d with $\eta \to 0$ as $\delta \to 0$. Since every one of these sets W satisfies $|W \setminus V| \geq |K|\delta d$, we have $\mathscr{M} = \bigcup_{W \in \mathscr{W}} \mathscr{M}_W$ and thus, combining with (3.2),

$$|\mathscr{M}| = \left| \bigcup_{W \in \mathscr{W}} \mathscr{M}_W \right| \leq |\mathscr{W}| q^{(1-\beta)d} \leq e^{\eta d} q^{(1-\beta)d}.$$

It follows that $h_\Sigma(\rho, \varepsilon, K, \delta) \leq \eta + (1 - \beta) \log q$, from which we conclude that $h_\Sigma(X, G) \leq (1 - \beta) \log q < \log q$. $\qquad\square$

3.7   Dualizing sofic measure entropy

Just as topological entropy (whether amenable or sofic) admits not only a formulation in terms of finite open covers but also a dual formulation in terms of pseudometrics and points in the space, one can express measure entropy in a dual way that similarly replaces partitions with points [34]. In fact this viewpoint brings measure entropy into close technical alignment with topological entropy, as the definitions will be identical except for an additional distributional requirement in the measure case. In particular we will need to work with topological models, i.e., we will assume that our p.m.p. action is actually a continuous action on a compact metrizable space with a G-invariant Borel probability measure. It is a well-known fact that every p.m.p. action is measure conjugate to such an action, and so there is no loss of generality in this setup.

Our setting is thus a continuous action $G \curvearrowright X$ of a countable sofic group on a compact metrizable space with a G-invariant Borel probability measure

108   *Recent Advances in Operator Theory and Operator Algebras*

μ. Let ρ be a continuous pseudometric on X, and recall the associated pseudometrics ρ_2 and ρ_∞ defined prior to Definition 3.6.1. The following is the same as Definition 3.6.1 except for the additional distributional requirement in (2). Here m denotes the uniform probability measure on $\{1, \ldots, d\}$.

Definition 3.7.1. *Let F be a finite subset of G, L a finite subset of $C(X)$, and $\delta > 0$. Let $\sigma : G \to \mathrm{Sym}(d)$ for some $d \in \mathbb{N}$. Define $\mathrm{Map}_\mu(\rho, F, L, \delta, \sigma)$ to be the set of all maps $\varphi : \{1, \ldots, d\} \to X$ such that*

1. $\rho_2(\varphi\sigma_s, \alpha_s\varphi) \le \delta$ for all $s \in F$, where α_s is the transformation $x \mapsto sx$ of X, and

2. $|(\varphi_* \mathrm{m})(f) - \mu(f)| \le \delta$ for all $f \in L$.

Definition 3.7.2. *We define*

$$h_{\Sigma,\mu}(\rho) = \sup_{\varepsilon>0} \inf_{F} \inf_{L} \inf_{\delta>0} \limsup_{i \to \infty} \frac{1}{d_i} \log N_\varepsilon(\mathrm{Map}_\mu(\rho, F, L, \delta, \sigma_i), \rho_\infty),$$

where F ranges over the finite subsets of G and L ranges over the finite subsets of $C(X)$.

Recall that ρ is *dynamically generating* if for all distinct $x, y \in X$ there is an $s \in G$ such that $\rho(sx, sy) > 0$.

Theorem 3.7.3. *Let ρ be a dynamically generating continuous pseudometric on X. Then $h_{\Sigma,\mu}(X, G) = h_{\Sigma,\mu}(\rho)$.*

The above formulation of measure entropy, in tandem with with Definition 3.6.4, is particularly useful for establishing the variational principle (Theorem 3.6.5). In particular, the fact that the topological entropy dominates the measure entropy with respect to every invariant Borel probability measure becomes a triviality.

Another approach due to Austin [2] uses what are called G-processes. These are right shift actions $G \curvearrowright (Y^G, \mu)$ where Y is a standard Borel space (e.g., a finite set) and μ is any Borel probability measure on Y^G. Every p.m.p. action $G \curvearrowright (X, \mu)$ is measure conjugate to a G-process via the conjugacy $X \to X^G$ sending x to $(sx)_{s \in G}$, with X^G equipped with the push-forward measure. Let $G \curvearrowright (Y^G, \mu)$ be a G-process, where Y is assumed to be a compact metric space with metric ρ, which induces the dynamically generating pseudometric $(x, y) \mapsto \rho(x_e, y_e)$ on Y^G. Let $\Sigma = \{\sigma_i : G \to \mathrm{Sym}(d_i)\}$ be a sofic approximation sequence. One can then compute the sofic entropy using Theorem 3.7.3. Austin's approach is to introduce a probability measure μ_i on Y^{d_i} for each i so that μ_i converges to μ in one of three possible (and in general nonequivalent) ways, which are referred to as *local weak* convergence*, *quenched convergence*, and *doubly quenched convergence*. The entropy is then defined by measuring the asymptotic exponential growth of the (ε, δ)-covering numbers of the measures μ_i with respect to the Hamming distances $(x, y) \mapsto$

$(1/d_i) \sum_{v=1}^{d_i} \rho(x_v, y_v)$ on Y^{d_i}, the (ε, δ)-covering number being the smallest cardinality of a set of δ-balls whose union has measure greater than ε. Using this definition Austin investigated the problem of additivity for sofic entropy undertaking product actions [2].

Abert and Weiss have also devised an alternative approach to sofic entropy that is particularly suited for handling Bernoulli actions [51].

3.8 Algebraic actions

While many of the influential applications of entropy theory for single transformations (e.g., to smooth dynamics) become inoperative when one passes to more general groups, there is one general class of action that is tailor-made for sofic entropy. These are the algebraic actions, and in particular the principal ones, which will be the main subject here. By an *algebraic action* we mean an action of a group on a compact Abelian group by (continuous) group automorphisms. For example, an $n \times n$ integer matrix A with determinant ± 1 induces an algebraic $\mathbb{Z}$-action on the n-torus $\mathbb{R}^n/\mathbb{Z}^n$ whose generating tranformation is given by left multiplication $x \mapsto Ax$ on column vectors.

Recall that the Pontrjagin dual $\widehat{A}$ of a locally compact Abelian group A is the locally compact Abelian group consisting of continuous homomorphisms from X to the circle $\mathbb{T} \cong \mathbb{R}/\mathbb{Z}$ with the topology of uniform convergence on compact subsets. When A is compact, the dual is discrete, and vice versa. The basic example is $\widehat{\mathbb{Z}} = \mathbb{T}$, and more generally $\widehat{\mathbb{Z}^n} = \mathbb{T}^n$. Pontrjagin duality asserts that A is isomorphic in a canonical way to its second dual. Every algebraic action $G \curvearrowright X$ induces an action $G \curvearrowright \widehat{X}$ by group automorphisms given by

$$(s\varphi)(x) = \varphi(s^{-1}x)$$

for $\varphi \in \widehat{X}$, $x \in X$, and $s \in G$. Given an action of G on a discrete Abelian group X, we can similarly induce an action of G on the compact Abelian group $\widehat{X}$ which inverts the above procedure, and so we get a bijective correspondence between algebraic actions of G and actions of G on discrete Abelian groups by group automorphisms. Moreover, the dual action $G \curvearrowright \widehat{X}$ naturally endows $\widehat{X}$ with the structure of a left module over the integral group ring $\mathbb{Z}G$ via the operation

$$\left(\sum_{s \in G} n_s s \right) \cdot \varphi = \prod_{s \in G} s\varphi^{n_s},$$

and one can also produce an algebraic action of G from a left $\mathbb{Z}G$-module in a natural way. This sets up a bijective correpondence between algebraic actions of G and left $\mathbb{Z}G$-modules and explains the use of the word "algebraic".

While only certain groups admit algebraic actions on, for example, the

n-torus (e.g., the matrix actions in the first paragraph), for an arbitrary G there is an abstract construction for conjuring a large class of algebraic actions out of the group itself via the integral group ring $\mathbb{Z}G$. As part of the module structure, the group G acts on $\mathbb{Z}G$ via shifting, and also more generally on $(\mathbb{Z}G)^n$ for a given $n \in \mathbb{N}$ by taking the product action. By regarding $(\mathbb{Z}G)^n$ as a discrete Abelian group, this gives us an algebraic action whose dual action $G \curvearrowright \widehat{(\mathbb{Z}G)^n}$ is the shift action $G \curvearrowright (\widehat{\mathbb{Z}^n})^G = (\mathbb{T}^n)^G$. Now take a matrix $A \in M_{k \times n}(\mathbb{Z}G)$. Then $(\mathbb{Z}G)^k A$ is a sub-$\mathbb{Z}G$-module of $(\mathbb{Z}G)^n$, and so the quotient $(\mathbb{Z}G)^n/(\mathbb{Z}G)^k A$ is a left $\mathbb{Z}G$-module, which like before we can consider as a discrete Abelian group on which G acts by shifting. The left $\mathbb{Z}G$-modules that arise this way are precisely the finitely presented left $\mathbb{Z}G$-modules. The dual action $G \curvearrowright \widehat{(\mathbb{Z}G)^n/(\mathbb{Z}G)^k A}$, which we will write simply as $G \curvearrowright X_A$, is a restriction of the full shift $G \curvearrowright (\widehat{\mathbb{T}^n})^G$.

In the square case $k = n$ and under an injectivity assumption on A, the entropy of the action $G \curvearrowright X_A$ can be expressed in terms of the Fuglede–Kadison determinant, whose definition we now review. If M is a unital C*-algebra with tracial state tr, we can extend this to a tracial positive linear functional on the C*-algebra $M_n(M)$ of $n \times n$ matrices over M by summing the traces of the diagonal elements. By the continuous functional calculus and the Riesz representation theorem, tr induces a Borel measure $\mu_{|A|}$ on the spectrum $\mathrm{spec}|A|$ of $|A|$ satisfying

$$\mathrm{tr}(f(|A|)) = \int_{\mathrm{spec}(|A|)} f(t) \, d\mu_{|A|}(t)$$

for $f \in C(\mathrm{spec}(|A|))$. The *Fuglede–Kadison determinant* of A (with respect to tr) is then defined by

$$\mathrm{det}_M A = e^{\int_{\mathrm{spec}(|A|)} \log t \, d\mu_{|A|}(t)}.$$

In the matrix algebra case $M = M_d$, for example, we have $\mathrm{det}_{M_d} A = |\det A|^{1/d}$ for $A \in M_d$, where det on the right is the ordinary determinant.

Here we are interested in the case where M is the group von Neumann algebra $\mathscr{L}G$ of G, which is the von Neumann algebra generated by the images of elements of G under the left regular representation $\lambda : G \to \mathscr{B}(\ell_2(G))$, with the canonical tracial state

$$\mathrm{tr}(a) = \langle a\delta_e, \delta_e \rangle,$$

where δ_e is the standard basis vector in $\ell_2(G)$ indexed by the identity element e of G. The following theorem of Hayes [29] is the culmination of a series of results by various authors addressing progressively broader classes of groups, starting with the work of Yuzvinskii for $G = \mathbb{Z}$ in the 1960s [52] and including Lind, Schmidt, and Ward's treatment of the case $G = \mathbb{Z}^n$ [41], where the Fuglede–Kadison determinant is the same as the Mahler measure, the expansion of the theory beyond the Abelian setting by Deninger in [12], and the work of Li [38] and Li and Thom [40] on the general amenable case.

Theorem 3.8.1. *Let $n \in \mathbb{N}$ and let A be an element of $M_n(\mathbb{Z}G)$ which is injective when regarded as a left operator on $\ell_2(G)^n$. Let Σ be a sofic approximation sequence for G and write μ for the Haar measure on X_A. Then*

$$h_\Sigma(X_A, G) = h_{\Sigma,\mu}(X_A, G) = \log \det_{\mathscr{L}G} A.$$

3.9   Further developments

Seward has undertaken a penetrating study of what he has called the *Rokhlin entropy* of an ergodic p.m.p. action $G \curvearrowright (X, \mu)$, defined as the infimum of the Shannon entropies of all generating countable Borel partitions [46, 47, 48]. This quantity coincides with amenable measure entropy when G is amenable, and is an upper bound for sofic measure entropy when G is sofic. Among many other things, Seward generalized Krieger's generator theorem by showing that for any finite distribution with Shannon entropy strictly less than the Rokhlin entropy there is a generating partition with this distribution.

The sofic entropy of algebraic actions and their extensions, with connections to cost and ℓ^2-Betti numbers, was studied by Gaboriau and Seward in [20]. The investigation of entropy and combinatorial independence for actions of amenable groups on compact spaces carried out by Kerr and Li in [32] was broadened to the sofic setting in [35]. In [39] Li developed a sofic version of mean dimension, a dimensional entropy-type invariant originally introduced by Gromov for actions of the integers and other amenable groups.

In [27] Hayes developed a powerful approach to sofic entropy using Polish models and applied it to the analysis of spectral properties and completely positive entropy. He also used it in [28] to compute the sofic entropy of Gaussian actions.

A notion of sofic dimension for discrete measured groupoids, and in particular for p.m.p. actions, was developed by Dykema, Kerr, and Pichot in [15]. The idea is to count all local models for the group and the action simultaneously (or more generally all local models for the groupoid) and measure the growth of this relative to the size of the full permutation groups of the finite sets in which the models reside. For free p.m.p. actions of free groups of finite rank, the dynamics washes out of the picture and one ends up with the rank of the group. As sofic dimension is an invariant of orbit equivalence, this gives another proof of Gaboriau's theorem that free p.m.p. actions of free groups of different finite ranks are never orbit equivalent.

Prior to his development of sofic entropy, Bowen introduced a notion of entropy for p.m.p. actions of free groups of finite rank using a formula involving Shannon entropy [4]. This quantity, called the f-invariant, was shown in [6] to be expressible as a limit over d of averages of local sofic entropies over all homomorphisms from the group into $\text{Sym}(d)$. When d is large such a homomorphism is, with high probability, a good sofic approximation (i.e., the approximate freeness condition holds in addition to the homomorphism property) .

References

[1] R. L. Adler, A. G. Konheim, and M. H. McAndrew. Topological entropy. *Trans. Amer. Math. Soc.* **114** (1965), 309–319.

[2] T. Austin. *Forum Math. Sigma* **4** (2016), e25, 79 pp.

[3] B. Blackadar and E. Kirchberg. Generalized inductive limits of finite-dimensional C^*-algebras. *Math. Ann.* **307** (1997), 343–380.

[4] L. Bowen. A measure-conjugacy invariant for free group actions. *Ann. Math. (2)* **171** (2010), 1387–1400.

[5] L. Bowen. Measure conjugacy invariants for actions of countable sofic groups. *J. Amer. Math. Soc.* **23** (2010), 217–245.

[6] L. Bowen. The ergodic theory of free group actions: entropy and the f-invariant. *Groups Geom. Dyn.* **4** (2010), 419–432.

[7] L. Bowen. Every countably infinite group is almost Ornstein. In: *Dynamical Systems and Group Actions*, 67–78, *Contemp. Math.*, 567, Amer. Math. Soc., Providence, RI, 2012.

[8] L. Bowen. Sofic entropy and amenable groups. *Ergodic Theory Dynam. Systems* **32** (2012), 427–466.

[9] R. Bowen. Entropy for group endomorphisms and homogeneous spaces. *Trans. Amer. Math. Soc.* **153** (1971), 401–414.

[10] N. P. Brown and N. Ozawa. C^*-Algebras and Finite-Dimensional Approximations. Graduate Studies in Mathematics, 88. American Mathematical Society, Providence, RI, 2008.

[11] P. de la Harpe. *Topics in Geometric Group Theory.* University of Chicago Press, Chicago, 2000.

[12] C. Deninger. Fuglede–Kadison determinants and entropy for actions of discrete amenable groups. *J. Amer. Math. Soc.* **19** (2006), 737–758.

[13] E. I. Dinaburg. A correlation between topological entropy and metric entropy. *Dokl. Akad. Nauk SSSR* **190** (1970), 19–22.

[14] E. I. Dinaburg. A connection between various entropy characterizations of dynamical systems. *Izv. Akad. Nauk SSSR Ser. Mat.* **35** (1971), 324–366.

114 *References*

[15] K. Dykema, D. Kerr, and M. Pichot. Sofic dimension for discrete measured groupoids. *Trans. Amer. Math. Soc.* **366** (2014), 707–748.

[16] G. Elek and E. Szabó. On sofic groups. *J. Group Theory* **9** (2006), 161–171.

[17] G. Elek and E. Szabó. Sofic representations of amenable groups. *Proc. Amer. Math. Soc.* **139** (2011), 4285–4291.

[18] G. Elliott, G. Gong, H. Lin, and Z. Niu. On the classification of simple amenable C*-algebras with finite decomposition rank, II. arXiv:1507.03437.

[19] D. Gaboriau. Entropie sofique. Séminaire Bourbaki, 2015–2016, no. 1108.

[20] D. Gaboriau and B. Seward. Cost, ℓ^2-Betti numbers and the sofic entropy of some algebraic actions. Preprint.

[21] G. Gong, H. Lin, and Z. Niu. Classification of finite simple amenable $\mathscr{Z}$-stable C*-algebras. arXiv:1501.00135.

[22] T. N. T. Goodman. Relating topological entropy and measure entropy. *Bull. London Math. Soc.* **3** (1971), 176–180.

[23] W. L. Goodwyn. Topological entropy bounds measure-theoretic entropy. *Proc. Amer. Math. Soc.* **23** (1969), 679–688.

[24] W. Gottschalk. Some general dynamical notions. In: *Recent Advances in Topological Dynamics (Proc. Conf. Topological Dynamics, Yale Univ., New Haven, Conn., 1972; in honor of Gustav Arnold Hedlund)*, 120–125. *Lecture Notes in Math.*, Vol. 318, Springer, Berlin, 1973.

[25] M. Gromov. Endomorphisms of symbolic algebraic varieties. *J. Eur. Math. Soc.* **1** (1999), 109–197.

[26] U. Haagerup and S. Thorbjørnsen. A new application of random matrices: $\text{Ext}(C^*_{\text{red}}(F_2))$ is not a group. *Ann. of Math. (2)* **162** (2005), 711–775.

[27] B. Hayes. Polish models and sofic antropy. To appear in *J. Inst. Math. Jussieu*.

[28] B. Hayes. Sofic entropy of Gaussian actions. To appear in *Ergodic Theory Dynam. Systems*.

[29] B. Hayes. Fuglede–Kadison determinants and sofic entropy. *Geom. Funct. Anal.* **26** (2016), 520–606.

[30] D. Kerr. Sofic measure entropy via finite partitions. *Groups Geom. Dyn.* **7** (2013), 617–632.

References 115

[31] D. Kerr. Bernoulli actions of sofic groups have completely positive entropy. *Israel J. Math.* **202** (2014), 461–474.

[32] D. Kerr and H. Li. Independence in topological and C*-dynamics. *Math. Ann.* **338** (2007), 869–926.

[33] D. Kerr and H. Li. Entropy and the variational principle for actions of sofic groups. *Invent. Math.* **186** (2011), 501–558.

[34] D. Kerr and H. Li. Soficity, amenability, and dynamical entropy. *Amer. J. Math.* **135** (2013), 721–761.

[35] D. Kerr and H. Li. Combinatorial independence and sofic entropy. *Commun. Math. Stat.* **1** (2013), 213–257.

[36] D. Kerr and H. Li. *Ergodic Theory: Independence and Dichotomies.* Springer Monographs in Mathematics series. Springer, 2016.

[37] J. C. Kieffer. A generalized Shannon–McMillan theorem for the action of an amenable group on a probability space. *Ann. Probability* **3** (1975), 1031–1037.

[38] H. Li. Compact group automorphisms, addition formulas and Fuglede–Kadison determinants. *Ann. Math. (2)* **176** (2012), 303–347.

[39] H. Li. Sofic mean dimension. *Adv. Math.* **244** (2013), 570–604.

[40] H. Li and A. Thom. Entropy, determinants, and L^2-torsion. *J. Amer. Math. Soc.* **27** (2014), 239–292.

[41] D. Lind, K. Schmidt, and T. Ward. Mahler measure and entropy for commuting automorphisms of compact groups. *Invent. Math.* **101** (1990), 593–629.

[42] H. Matui and Y. Sato. Decomposition rank of UHF-absorbing C*-algebras. *Duke Math. J.* **163** (2014), 2687–2708.

[43] J. Moulin Ollagnier. *Ergodic Theory and Statistical Mechanics.* Lecture Notes in Mathematics, 1115. Springer-Verlag, Berlin, 1985.

[44] D. Ornstein. Bernoulli shifts with the same entropy are isomorphic. *Advances in Math.* **4** (1970), 337–352.

[45] D. S. Ornstein and B. Weiss. Entropy and isomorphism theorems for actions of amenable groups. *J. Analyse Math.* **48** (1987), 1–141.

[46] B. Seward. Weak containment and Rokhlin entropy. Preprint.

[47] B. Seward. Krieger's finite generator theorem for ergodic actions of countable groups II. Preprint.

References

[48] B. Seward. Krieger's finite generator theorem for ergodic actions of countable groups I. Preprint.

[49] A. Tikuisis, S. White, and W. Winter. Quasidiagonality of nuclear C^*-algebras. *Ann. Math. (2)* **185** (2017), 229–284.

[50] B. Weiss. Sofic groups and dynamical systems. In: *Ergodic Theory and Harmonic Analysis (Mumbai, 1999)*. Sankhya Ser. A **62** (2000), 350–359.

[51] B. Weiss. Entropy and actions of sofic groups. *Discrete Contin. Dyn. Syst. Ser. B* **20** (2015), 3375–3383.

[52] S. A. Yuzvinskii. Calculation of the entropy of a group-endomorphism. (Russian) *Sibirsk. Mat. Ž.* **8** (1967), 230–239.

[53] G. Zhang. Local variational principle concerning entropy of sofic group action. *J. Funct. Anal.* **262** (2012), 1954–1985.

Chapter 4

The solution of the Kadison–Singer problem

Dan Timotin

4.1	Introduction	117
4.2	The Kadison–Singer problem	119
	4.2.1 Pure states	119
	4.2.2 The Kadison–Singer conjecture	120
	4.2.3 The paving conjecture	121
4.3	Intermezzo: what we will do next and why	123
	4.3.1 General plan	123
	4.3.2 Sketch of the proof	123
4.4	Analytic functions and univariate polynomials	124
	4.4.1 Preliminaries	125
	4.4.2 Nice families	126
4.5	Several variables: real stable polynomials	129
	4.5.1 General facts	129
	4.5.2 The barrier function	131
4.6	Characteristic and mixed characteristic polynomials	135
	4.6.1 Mixed characteristic polynomial	135
	4.6.2 Decomposing in rank one matrices and the characteristic polynomial	136
4.7	Randomization	138
	4.7.1 Random matrices and determinants	138
	4.7.2 Probability and partitions	140
4.8	Proof of the paving conjecture	141
4.9	Final remarks	144

Abstract. In the summer of 2013 Marcus, Spielman, and Srivastava gave a surprising and beautiful solution to the Kadison–Singer problem. The current presentation is slightly more didactical than other versions that have appeared since; it hopes to contribute to a thorough understanding of this amazing proof.

4.1 Introduction

The Kadison–Singer problem posed in [7] in the 1950s, probably in relation to a statement of Dirac concerning the foundations of quantum mechanics. It has since acquired a life of its own. On one hand, there have been several notable attempts to prove it. On the other hand, it has been shown that it is equivalent to various problems in Hilbert space theory, frame theory, geometry of Banach spaces, etc. However, for five decades the problem has remained unsolved.

It is therefore very remarkable that in 2013 a proof was given by Marcus, Spielman, and Srivastava in [10]. The methods used were rather unexpected; moreover, they had shown their strength in some totally unrelated areas (Ramanujan graphs). They also have a very elementary flavor: most of the proof is based on a delicate analysis of the behavior of polynomials in one or several variables.

In the time that has passed since a better grasp of the proof was achieved, most notably through Terence Tao's entry in his blog [14] (but see also [11]), still remains an astonishing piece of research, obtaining spectacular results on a long-standing conjecture through some not very complicated and apparently unrelated arguments.

The purpose of these notes is to contribute toward a better understanding of the MSS proof. There is of course no pretense to any originality: the content is essentially in [10], with some supplementary simplification due to [14] (and occasionally to [15]). But we have tried to make it easier to follow by separating clearly the different steps and emphasizing the main arguments; also, in various places we have gone into more detail than in the other presentations. It is to be expected that the methods of [10] might lead to new fruitful applications, and so it seemed worthwhile to analyze them in detail.

It is clear from the above that the notes concentrate on the MMS proof, so there will be very little about the Kadison–Singer problem itself and about the plethora of research that had evolved in the last fifty years on its relations to other domains. In particular, with one exception that we need to use (the paving conjecture), we will not discuss the different reformulations and equivalent statements that have been obtained. For all these matters, one may consult former beautiful presentations, as for instance [4].

We will give in the next section a brief presentation of the original problem, as well as of another assertion, the paving conjecture, which was shown soon afterwards to imply it. The description of the remaining part of the paper is postponed until Section 4.3, where the reader will have a general overview of the development of the proof.

These notes have been written for a series of lectures given in December 2014 at the Indian Statistical Institute in Bangalore, in the framework of the meeting *Recent Advances in Operator Theory and Operator Algebras*. We

The solution of the Kadison–Singer problem 119

thank B.V.R. Bhat, J. Sarkar, and V.S. Sunder for the excellent work done in organizing the workshop and the conference, as well as for the invitation to present the lectures.

4.2 The Kadison–Singer problem

4.2.1 Pure states

The material in this subsection is contained in standard books on C^*-algebras (see, for instance, [6]).

We denote by $B(\mathcal{H})$ the algebra of all bounded linear operators on the Hilbert space $\mathcal{H}$. A C^*-algebra $\mathcal{A} \subset B(\mathcal{H})$ is a norm closed subalgebra of $B(\mathcal{H})$, closed to the operation of taking the adjoint, and containing the identity.

A *state* on a C^*-algebra $\mathcal{A}$ is a linear continuous map $\varphi : \mathcal{A} \to \mathbb{C}$, which is positive (meaning that $\varphi(a^*a) \geq 0$ for any $a \in \mathcal{A}$), and such that $\varphi(I) = 1$. One proves then that $\|\varphi\| = 1$ and that φ satisfies the *Cauchy–Schwarz* inequality

$$|\varphi(b^*a)|^2 \leq \varphi(a^*a)\varphi(b^*b) \tag{4.1}$$

for all $a, b \in \mathcal{A}$.

The set $\mathfrak{S}(\mathcal{A})$ of all states on $\mathcal{A}$ is a convex, w^*-compact subset of $\mathcal{A}^*$. A state φ is called *pure* if it is an extreme point of $\mathfrak{S}(\mathcal{A})$.

Example 4.2.1. *If $\mathcal{A}$ is commutative, then by Gelfand's theorem it is isomorphic to $C(X)$, the algebra of continuous functions on the compact space X of all* characters *(multiplicative homomorphisms) $\chi : \mathcal{A} \to \mathbb{C}$. The dual $C(X)^*$ is formed by all Borel measures on X, and $\mathfrak{S}(C(X))$ is the set of probability measures on X. Pure states are precisely Dirac measures. In particular (and this is a fact that we will use below) a pure state on a commutative C^*-algebra is multiplicative.*

Example 4.2.2. *If $\mathcal{A} = B(\mathcal{H})$, $\xi \in \mathcal{H}$, and $\|\xi\| = 1$, then one can prove that $\varphi_\xi(T) := \langle T\xi, \xi \rangle$ is a pure state. This fact will not be used in the sequel.*

By a theorem of Krein (see, for instance, [12, Ch.I.10]) any state φ on a C^*-algebra $\mathcal{A}$ extends to a state $\tilde{\varphi}$ on $B(\mathcal{H})$. The set K_φ of all extensions of φ is a convex w^*-compact subset of $B(\mathcal{H})^*$.

Lemma 4.2.3. *If φ is a pure state on $\mathcal{A} \subset B(\mathcal{H})$, then the extreme points of K_φ are pure states of $B(\mathcal{H})$.*

Proof. Suppose φ is an extreme point of K_φ. If $\tilde{\varphi} = \frac{1}{2}(\psi_1 + \psi_2)$, with ψ_1, ψ_2 states on $B(\mathcal{H})$, then $\varphi = \frac{1}{2}(\psi_1|\mathcal{A} + \psi_2|\mathcal{A})$. Since φ is pure, we must have $\psi_1|\mathcal{A} = \psi_2|\mathcal{A} = \varphi$, so $\psi_1, \psi_2 \in K_\varphi$, and therefore $\psi_1 = \psi_2 = \varphi$. $\square$

120　　*Recent Advances in Operator Theory and Operator Algebras*

Consequently, a pure state φ on $\mathcal{A}$ has a unique extension to a state on $B(\mathcal{H})$ if and only if it has a unique *pure* extension to a state on $B(\mathcal{H})$.

4.2.2　The Kadison–Singer conjecture

From now on we will suppose that the Hilbert space $\mathcal{H}$ is $\ell^2 = \ell^2(\mathbb{N})$ and we will consider matrix representations of operators in $B(\ell^2)$ with respect to the usual canonical basis of ℓ^2. We define $\mathcal{D}$ to be the C^*-algebra of operators on ℓ^2 whose matrix is diagonal. Note that the map diag $: B(\ell^2) \to \mathcal{D}$ which sends an operator T to the diagonal operator having the same diagonal entries is continuous, positive, of norm 1.

We may now state the *Kadison–Singer problem*:

Does any pure state on $\mathcal{D}$ extend uniquely to a state on $B(\ell^2)$?

Although Kadison and Singer originally thought a negative answer to this question as more probable, in view of its eventual positive answer we will state the conjecture in the affirmative form.

Kadison–Singer Conjecture (KS). Any pure state on $\mathcal{D}$ extends uniquely to a state on $B(\ell^2)$.

The first thing to note is that any state $\varphi \in \mathfrak{S}(\mathcal{D})$ has a "canonical" extension to $\mathfrak{S}(B(\ell^2))$, given by

$$\tilde{\varphi}(T) = \varphi(\mathrm{diag}(T)). \tag{4.1}$$

So the problem becomes whether $\tilde{\varphi}$ is or not the unique extension of φ to $B(\ell^2)$. If ψ is another extension of φ and $T \in B(\ell^2)$, then

$$\psi(T - \mathrm{diag}\, T) = \psi(T) - \varphi(\mathrm{diag}\, T) = \psi(T) - \tilde{\varphi}(T).$$

So $\psi = \tilde{\varphi}$ if and only if $\psi(T - \mathrm{diag}\, T) = 0$ for any $T \in B(\ell^2)$, which is equivalent to say that $\psi(T) = 0$ for any $T \in B(\ell^2)$ with $\mathrm{diag}\, T = 0$. As a consequence, we have the following simple lemma.

Lemma 4.2.4. *(KS) is true if and only if any extension $\psi \in \mathfrak{S}(B(\ell^2))$ of a pure state on $\mathcal{A}$ satisfies*

$$\mathrm{diag}\, T = 0 \implies \psi(T) = 0.$$

In fact, pure states of $\mathcal{D}$ can be described more precisely. Indeed, being a commutative algebra, $\mathcal{D}$ is isomorphic to $C(X)$ (as noted in Example 4.2.1). One can identify X precisely: it is $\beta\mathbb{N}$, the Stone-Cech compactification of $\mathbb{N}$. We do not need this fact, but will use only a simple observation.

Lemma 4.2.5. *If φ is a pure state on $\mathcal{D}$ and $P \in \mathcal{D}$ is a projection, then $\varphi(P)$ is either 0 or 1.*

The solution of the Kadison–Singer problem 121

Proof. It has been noted above (see Example 4.2.1) that φ is multiplicative. Then $\varphi(P) = \varphi(P^2) = \varphi(P)^2$, whence $\varphi(P)$ is either 0 or 1. $\qquad\square$

Remark 4.2.6. *As hinted in the introduction, although in the original paper [7] there is no mention of quantum mechanics, in subsequent papers the authors state as source for the problem the work of Dirac on the foundation of quantum mechanics [5]. For some comments on this, see Section 4.9.1 below.*

4.2.3   The paving conjecture

Instead of dealing directly with the Kadison–Singer conjecture, we intend to prove a statement about finite-dimensional matrices, which is usually known as Anderson's *paving conjecture* [1]. We use the notation $\mathcal{D}_m$ to indicate diagonal $m \times m$ matrices and diag_m the corresponding map from $M_m(\mathbb{C})$ to $\mathcal{D}_m$.

Paving Conjecture (PC). For any $\epsilon > 0$ there exists $r \in \mathbb{N}$ such that the following is true.

For any $m \in \mathbb{N}$ and $T \in B(\mathbb{C}^m)$ with $\mathrm{diag}_m T = 0$, there exist projections $Q_1, \ldots, Q_r \in \mathcal{D}_m$, with $\sum_{i=1}^r Q_i = I_m$, and

$$\|Q_i T Q_i\| \leq \epsilon \|T\|$$

for all $i = 1, \ldots, r$.

A diagonal projection $Q \in \mathcal{D}_m$ has its entries 1 or 0, so it is defined by a subset $S \subset \{1, \ldots, m\}$. Thus diagonal projections $Q_1, \ldots Q_r \in \mathcal{D}_m$ with $\sum_{i=1}^r Q_i = I_m$ correspond to partitions $\{1, \ldots, m\} = S_1 \cup \cdots \cup S_r$, $S_i \cap S_j = \emptyset$ for $i \neq j$.

It is important that in the statement of (PC) the number r does not depend on m. This allows us to deduce from (PC) a similar statement, in which $\mathbb{C}^m$ is replaced with the whole ℓ^2, is also true. We formulate this as a lemma.

Lemma 4.2.7. *If (PC) is true, then for any $\epsilon > 0$ there exists $r \in \mathbb{N}$ such that, for any $T \in B(\ell^2)$ with $\mathrm{diag}\, T = 0$ one can find projections $Q_1, \ldots, Q_r \in \mathcal{D}$, with $\sum_{i=1}^r Q_i = I$, and*

$$\|Q_i T Q_i\| \leq \epsilon \|T\|$$

for all $i = 1, \ldots, r$.

Proof. Embed $\mathbb{C}^m$ canonically into ℓ^2 on the first m coordinates and denote by E_m the corresponding orthogonal projection. For $T \in B(\ell^2)$ denote $T_m = E_m T E_m$. Applying (PC), one finds diagonal projections $Q_1^{(m)}, \ldots, Q_r^{(m)}$, such that $\sum_{i=1}^r Q_i^{(m)} = I_m$ and $\|Q_i^{(m)} T_m Q_i^{(m)}\| \leq \epsilon \|T_m\|$.

Now, diagonal projections in $B(\ell^2)$ can be identified with subsets of $\mathbb{N}$, and therefore with elements in the compact space $\{0, 1\}^{\mathbb{N}}$. In this compact space any sequence has a convergent subsequence; therefore a diagonal argument

122   *Recent Advances in Operator Theory and Operator Algebras*

will produce an increasing subsequence of positive integers m_k, such that for each $i = 1, \ldots, r$ we have $Q_i^{(m_k)} \to Q_i$ for some Q_i. We have

$$\sum_{i=1}^{r} Q_i = \lim_{k \to \infty} \sum_{i=1}^{r} Q_i^{(m_k)} = \lim_{k \to \infty} I_{m_k} = I.$$

If $\xi, \eta \in \ell^2$ are vectors with finite support, then $\xi, \eta \in \mathbb{C}^{d_k}$ for some k, and then

$$|\langle Q_i T Q_i \xi, \eta \rangle| = |\langle T Q_i \xi, Q_i \eta \rangle| = |\langle T_{m_k} Q_i^{(m_k)} \xi, Q_i^{(m_k)} \eta \rangle|$$
$$= |\langle Q_i^{(m_k)} T_{m_k} Q_i^{(m_k)} \xi, \eta \rangle|$$
$$\leq \|Q_i^{(m_k)} T_{m_k} Q_i^{(m_k)}\| \|\xi\| \|\eta\| \leq \epsilon \|T\| \|\xi\| \|\eta\|. \qquad \square$$

The paving conjecture is actually equivalent to the Kadison–Singer conjecture, but we will need (and prove) only one of the implications.

Proposition 4.2.8. *The Paving Conjecture implies the Kadison–Singer Conjecture.*

Proof. Fix $\epsilon > 0$, and suppose that r satisfies the conclusion of Lemma 4.2.7. Take a pure state $\psi \in \mathfrak{S}(B(\ell^2))$ and an operator $T \in B(\ell^2)$ with $\operatorname{diag} T = 0$. By Lemma 4.2.4 we have to show that $\psi(T) = 0$.

Let Q_i be the diagonal projections associated to T by Lemma 4.2.7. By Lemma 4.2.5, $\psi(Q_i) = \varphi(Q_i)$ is 0 or 1 for each i. Since $1 = \varphi(I) = \sum_{i=1}^{r} \varphi(Q_i)$, it follows that there exists some i_0 for which $\varphi(Q_{i_0}) = 1$, while $\varphi(Q_i) = 0$ for $i \neq i_0$.

If $i \neq i_0$, then (4.1) implies

$$|\psi(Q_i R)| \leq \psi(Q_i^* Q_i) \psi(R^* R) = \psi(Q_i) \psi(R^* R) = 0,$$

and similarly $\psi(R Q_i) = 0$ for all $R \in B(\ell^2)$. Therefore

$$\psi(T) = \sum_{i=1}^{r} \sum_{j=1}^{r} \psi(Q_i T Q_j) = \psi(Q_{i_0} T Q_{i_0}).$$

But the projections Q_i have been chosen such as to have $\|Q_{i_0} T Q_{i_0}\| \leq \epsilon \|T\|$, so

$$|\psi(T)| \leq \|Q_{i_0} T Q_{i_0}\| \leq \epsilon \|T\|.$$

Since this is true for any $\epsilon > 0$, it follows that $\psi(T) = 0$, and the proposition is proved. $\qquad \square$

4.3 Intermezzo: what we will do next and why

4.3.1 General plan

As noted above, we intend to prove the Paving Conjecture. The proof will lead us on an unexpected path, so we explain here its main steps.

The Paving Conjecture asks us to find, for a given matrix T, diagonal projections Q_i that achieve certain norm estimates (namely, $\|Q_i T Q_i\| \leq \epsilon\|T\|$). Among the different ways to estimate the norm, the proof in [10] chooses a rather unusual one: it uses the fact that the norm of a positive operator is its largest eigenvalue. So we have to consider characteristic polynomials of matrices—in fact, the largest part of the proof is dedicated to estimating roots of such polynomials. (Although it has nothing to do with (KS), one should note the added benefit that we find a way to control with no extra effort all eigenvalues of the matrix, not only the largest one.)

On the other hand, to achieve this control we need to make an unexpected detour: though the characteristic polynomial depends on a single variable, in order to control it one has to go through multivariable polynomials and to use the theory of real stable polynomials as developed by Borcea and Brändén [2]. This may seem unnatural, but it should be mentioned that Borcea and Brändén have already obtained through their methods spectacular results, in particular solving long-standing conjectures in matrix theory that also seemed at first sight to involve just a single complex variable [2, 3]. So maybe one should not be so surprised after all.

A second feature of the proof is its use, at some point, of a random space. After obtaining certain results about eigenvalues of usual matrices, suddenly random matrices appear on the scene. In fact, the use of randomness is not really essential; it rather provides a convenient notation for computing averages. As noted in the previous section, to prove (PC) we need to find a partition of a finite set $\{1, \ldots, m\}$ into r subsets with certain properties. The random space eventually considered is finite; its elements are all different partitions, and no subtle probability is used: all decompositions are assumed to be equally probable. What we will achieve eventually is an estimate on the average of the largest eigenvalue, which will lead to an individual estimate for at least one point of the random space—that is, for one partition. This will be the desired partition.

4.3.2 Sketch of the proof

We summarize here the development of the proof. As announced above, we intend to discuss the eigenvalues of positive matrices, which are roots of the characteristic polynomial. So we need some preparation concerning polynomials and their roots; this is done first in one variable in Section 4.4. The main result here is Theorem 4.4.9, that shows that certain families of polynomials have roots that behave unexpectedly well with respect to averages. This will be used in Section 4.7 to link eigenvalues of random matrices to their averages.

But we have to go to polynomials in several variables, namely real stable polynomials, which are defined by a condition on their roots. Section 4.5 is dedicated to real stable polynomials; after presenting their main properties, we are especially interested in some delicate estimate on the location of the roots, which is done through an associated function called the barrier function. The properties of the barrier function represent the most technical and not very transparent part of the proof. The main thing to be used in the sequel is Theorem 4.5.8, that puts some restriction on the roots of a real stable polynomial.

We apply these facts to characteristic polynomials in Section 4.6. The voyage through several variables done for polynomials has a correspondent here in the introduction of the *mixed characteristic polynomial*, which depends on several matrices. It happens to be the restriction to one variable of a real stable polynomial, and so Theorem 4.5.8 can be used in Theorem 4.6.4 to bound the roots of a mixed characteristic polynomial. Further, this bound translates in a bound for a usual characteristic polynomial in the particular case when the matrices have rank one, since then the mixed characteristic polynomial is precisely the characteristic polynomial of their sum.

Section 4.7 introduces random matrices; as discussed above, the probability space in view is that of all possible partitions. The main result, Theorem 4.7.2, uses the results of Section 4.4 to show that for a sum of independent random matrices of rank one, the eigenvalues of its average yield estimates for the averages of its eigenvalues, and thus for the eigenvalues of at least one point of the probability space. In particular, applying this fact in conjunction with the bound on eigenvalues obtained in Section 4.6, we will obtain a partition with certain norm properties in Theorem 4.7.5.

Finally, this last fact is put to good use in Section 4.8 to obtain a proof of the paving conjecture. The first step, that uses Theorem 4.7.5, obtains for orthogonal projections a quantitative version of (PC). To go from there to general operators has been well known for several decades and may be done in different ways. Here we use a dilation argument taken from [15] to obtain the paving conjecture for selfadjoint matrices; passing to general matrices is then immediate.

4.4 Analytic functions and univariate polynomials

4.4.1 Preliminaries

The following theorem in complex function theory is a consequence of Cauchy's argument principle.

Theorem 4.4.1. *Suppose (f_n) is a sequence of analytic functions on a domain $D \subset \mathbb{C}$, which converges uniformly on compacts to the function $f \not\equiv 0$. If Γ is a simple contour contained in D such that f has no zeros on Γ, then there is $n_0 \in \mathbb{N}$ such for $n \geq n_0$ the number of zeros of f_n and of f in the interior of Γ coincide.*

The next corollary is usually called Hurwitz's theorem if $m = 1$. The general case follows simply by induction (exercise!).

Corollary 4.4.2. *Suppose $p_n(z_1, \ldots, z_m)$ are polynomials in m variables, such that $p_n \to p$ uniformly on compacts in some domain $D \subset \mathbb{C}^m$. If p_m has no zeros in D for all m, then either p is identically zero, or it has no zeros in D.*

If f is a polynomial of degree n with all coefficients and all roots real, we denote its roots by

$$\rho_n(f) \leq \cdots \leq \rho_1(f).$$

Corollary 4.4.3. *Suppose $p_s(z) = \sum_{i=1}^{n} a_i(s)z^i$, with $a_i : I \to \mathbb{R}$ continuous functions on an interval $I \subset \mathbb{R}$, $a_n(s) \neq 0$ on I. If p_s has real roots for all $s \in I$, then the roots $\rho_1(p_s), \ldots, \rho_n(p_s)$ are continuous functions of $s \in I$.*

Proof. We use induction with respect to n. The case $n = 1$ is obvious. Then, for a general n, we prove first that $\rho_1(p_s)$ is continuous, say in $s_0 \in I$. Take $\epsilon > 0$, and suppose also that $p_{s_0}(s_0 \pm \epsilon) \neq 0$. By continuity of a_i, $p_s(s_0 \pm \epsilon) \neq 0$ for s sufficiently close to s_0, and so $p_s(z) \neq 0$ for z on the circle Γ of diameter $[s_0 - \epsilon, s_0 + \epsilon]$ (since all p_s have real roots). By Theorem 4.4.1 all p_s have at least one root inside Γ for s sufficiently close to s_0. A similar argument, using a circle at the right of $s_0 + \epsilon$, shows that the p_s have no roots larger than b. It follows that $\rho_1(p_s) \in (a, b)$ for s close to s_0.

If we write now $p_s(z) = (z - \rho_1(p_s))q_s(z)$, then q_s has degree $n - 1$ and continuous coefficients, so its roots are continuous by the induction hypothesis. But we have $\rho_i(p) = \rho_{i-1}(q)$ for $i \geq 2$. $\qquad\square$

Remark 4.4.4. *Even without the assumption that the roots are real, one can prove that there exist continuous functions $\rho_i(s) : I \to \mathbb{C}$, $i = 1, \ldots, n$, such that the roots of p_s are $\rho_1(s), \ldots, \rho_n(s)$ for all $s \in I$. The proof is more involved; see, for instance, [8, II.5.2].*

We prove the next two lemmas about polynomials with real coefficients and real roots.

Lemma 4.4.5. *Suppose the polynomial p of degree n has real coefficients, real roots, and the leading term positive. Moreover, assume that there exist real numbers $a_{n+1} < a_n < \cdots < a_1$ such that $\rho_j(p) \in [a_{j+1}, a_j]$ for all $j = 1, \ldots, n$. Then $(-1)^{j-1}p(a_j) \geq 0$ for all $j = 1, \ldots, n$.*

126   *Recent Advances in Operator Theory and Operator Algebras*

In other words, p changes signs (not necessarily strictly) on each of the intervals $[a_{j+1}, a_j]$.

Proof. We will use induction with respect to n. For $n = 1$ the claim is obviously true. Suppose it is true up to $n - 1$, and let p be a polynomial of degree n as in the statement of the lemma. There are two cases to consider.

Suppose first that the roots of p are exactly all points a_j except some a_{j_0}. Then p has only simple roots, so it changes signs in each of them. As $p(x) > 0$ for $x > a_1$, we have $p(x) < 0$ on (a_2, a_1), etc., up to $(-1)^{j_0-1} p(x) > 0$ on (a_{j_0+1}, a_{j_0-1}). Therefore $(-1)^{j_0-1} p(a_{j_0}) > 0$; the other inequalities are trivial.

In the remaining case, there is at least one root α of p that is not among the points a_j; suppose $\alpha \in (a_{j_0}, a_{j_0-1})$. If $p(z) = (z-\alpha)q(z)$, then q has degree $n - 1$ and satisfies the hypotheses of the lemma with respect to the points a_j with $j \neq 0$. Then $p(a_j)$ has the same sign as $q(a_j)$ for $j < j_0$ and opposite sign for $j > j_0$; from here it follows easily that the correct signs for q (which we know true by the induction hypothesis) produce the correct signs for p. $\square$

Lemma 4.4.6. *Suppose the polynomial p has real coefficients and all roots real. Then*

$$(-1)^k \left(\frac{d}{dx} \right)^k \frac{p'}{p}(x) > 0$$

for all $k \in \mathbb{N}$ and $x > \rho_1(p)$.

In particular, $\frac{p'}{p}$ is positive, nonincreasing, and convex for $x > \rho_1(p)$.

Proof. If $p(z) = \prod_{i=1}^n (z - \rho_i(p))$, then $\frac{p'}{p}(z) = \sum_{i=1}^n \frac{1}{z - \rho_i(p)}$, and

$$(-1)^k \left(\frac{d}{dx} \right)^k \frac{p'}{p}(x) = k! \sum_{i=1}^n \frac{1}{(z - \rho_i(p))^{k+1}}.$$

All terms in the last sum are positive for $x > \rho_1(p)$, so the lemma is proved. $\square$

4.4.2   Nice families

Suppose $\mathcal{F} = \{f_1, \ldots f_m\}$ is a family of polynomials of the same degree n. We denote

$$\rho_j^+(\mathcal{F}) := \max_{1 \leq i \leq m} \rho_j(f_i), \quad \rho_j^-(\mathcal{F}) := \min_{1 \leq i \leq m} \rho_j(f_i)$$

Definition 4.4.7. *A family of polynomials $\mathcal{F} = \{f_1, \ldots f_m\}$ of the same degree n is called a* nice family *iff:*

1. *the coefficient of the dominant term of every f_j is positive;*

2. *every f_j has all roots real;*

The solution of the Kadison–Singer problem   127

3. for all $j = 2, \ldots, n$ we have

$$\rho_j^+(\mathcal{F}) \le \rho_{j-1}^-(\mathcal{F}). \tag{4.1}$$

The usual formulation (including [9, 10]) is that the f_is *have a common interlacing.* Since the actual interlacing polynomial never enters our picture, we prefer this simpler phrasing.

Lemma 4.4.8. *(i) $\{f_1, \ldots f_m\}$ is nice iff every pair $\{f_r, f_s\}$, $r \ne s$, is nice.*

(ii) Every subfamily of a nice family is nice.

(iii) If $a \in \mathbb{R}$, then $\mathcal{F} = \{f_1, \ldots f_m\}$ is nice if and only if $\mathcal{G} = \{(x - a)f_1, \ldots (x - a)f_m\}$ is nice.

Proof. (i) and (ii) are immediate. For (iii), there are several cases to consider:

1. If $a \in [\rho_{j_0}^-(\mathcal{F}), \rho_{j_0}^+(\mathcal{F})]$ for some j_0, then

$$\rho_j^\pm(\mathcal{G}) = \rho_j^\pm(\mathcal{F}) \text{ for } j < j_0,$$
$$\rho_{j_0}^+(\mathcal{G}) = \rho_{j_0}^+(\mathcal{F}), \quad \rho_{j_0}^-(\mathcal{G}) = \rho_{j_0+1}^+(\mathcal{G}) = 1, \quad \rho_{j_0+1}^-(\mathcal{G}) = \rho_{j_0}^-(\mathcal{F}),$$
$$\rho_j^\pm(\mathcal{G}) = \rho_{j-1}^\pm(\mathcal{F}) \text{ for } j > j_0 + 1.$$

2. If $a \in (\rho_{j_0}^+(\mathcal{F}), \rho_{j_0-1}^-(\mathcal{F}))$ for some j_0, then

$$\rho_j^\pm(\mathcal{G}) = \rho_j^\pm(\mathcal{F}) \text{ for } j < j_0,$$
$$\rho_{j_0}^\pm(\mathcal{G}) = a,$$
$$\rho_j^\pm(\mathcal{G}) = \rho_{j-1}^\pm(\mathcal{F}) \text{ for } j > j_0.$$

The formulas in (1) are also valid if $a > \rho_1^+(\mathcal{F})$ (taking $j_0 = 1$) or $a < \rho_n^-(\mathcal{F})$ (taking $j_0 = n+1$). In all these cases one can easily check that (iii) is true. $\square$

As a consequence of Lemma 4.4.8, in order to check that a family is nice we can always assume that it has no common zeros.

The main theorem of this section is the characterization of nice families that follows.

Theorem 4.4.9. *Suppose $f_1, \ldots, f_m$ are all polynomials of degree n, with positive dominant coefficients. The following are equivalent:*

1. $\mathcal{F} = \{f_1, \ldots f_m\}$ is a nice family.

2. Any convex combination of $f_1, \ldots, f_m$ has only real roots.

If these conditions are satisfied, then for any $j = 1, \ldots, n$ we have

$$\min_i \rho_j(f_i) \le \rho_j(f) \le \max_i \rho_j(f_i) \tag{4.2}$$

for any convex combination $f = \sum_k t_k f_k$.

128   *Recent Advances in Operator Theory and Operator Algebras*

Proof. $(1) \implies (2)$. We may suppose by (ii) and (iii) of Lemma 4.4.8 that all coefficients t_k are positive and that the family has no common zeros. In particular, if we denote $\rho_j^\pm = \rho_j^\pm(\mathcal{F})$, this implies $\rho_j^- < \rho_j^+ \le \rho_{j-1}^-$ for all j.

We will apply Lemma 4.4.5 to each of the polynomials f_i and the points $\rho_n^- < \rho_{n-1}^- < \cdots < \rho_1^- < \rho_1^+$. We obtain then, for each $i = 1, \ldots, m$, that $(-1)^j f_i(\rho_j^-) \ge 0$ for all j, and $f_i(\rho_1^+) \ge 0$.

Fix j; since the family $\mathcal{F}$ has no common zero, at least one of f_i is nonzero in ρ_j^-, and so $(-1)^j f(\rho_j^-) > 0$. Similarly, $f(\rho_1^+) > 0$. Therefore on each of the intervals (ρ_j^-, ρ_{j-1}^-), as well as on (ρ_1^-, ρ_1^+), f changes sign (strictly), and therefore must have a root in the interior. Since there are n intervals, we have thus found n roots of f, and so all its roots are real. Moreover, we have obtained $\rho_j(f) > \rho_j^-$ for all j.

On the other hand, we might have used, in applying Lemma 4.4.5 to the polynomials f_i, the points $\rho_n^- < \rho_n^+ < \rho_{n-1}^+ \cdots < \rho_1^+$ instead of $\rho_n^- < \rho_{n-1}^- < \cdots < \rho_1^- < \rho_1^+$. A similar argument yields then $\rho_j(f) < \rho_j^+$ for all j. Therefore the inequalities (4.2) are proved.

$(2) \implies (1)$. According to Lemma 4.4.8 it is enough to prove the implication for two functions f_1, f_2, and we may also suppose that they have no common roots. Fix $2 \le j \le n$; we have to prove that $\rho_j^+ \le \rho_{j-1}^-$. Denote $f_t = t f_1 + (1 - t) f_2$ $(0 \le t \le 1)$. By Corollary 4.4.3 the function $t \mapsto \rho_j(f_t)$ is continuous on $[0, 1]$ and takes only real values; so its values for $0 < t < 1$ cover the interval (ρ_j^-, ρ_j^+). It follows that this interval cannot contain a root of either f_1 or f_2, since a common root of, say, f_1 and f_t is also a root of f_2.

Suppose then first that f_1 and f_2 have only simple roots. Then the intervals $[\rho_j^-, \rho_j^+]$ and $[\rho_{j-1}^-, \rho_{j-1}^+]$ have all four endpoints disjoint, and by definition $\rho_j^- < \rho_{j-1}^-$. If $\rho_{j-1}^- \in (\rho_j^-, \rho_j^+)$, this would contradict the conclusion of the preceding paragraph. So $\rho_{j-1}^- > \rho_j^+$ and (4.1) is proved.

To obtain the general case, note first that f_t has all roots simple for $0 < t < 1$. Indeed, a multiple solution x of $f_t = 0$ would also be a multiple solution of $\frac{f_2}{f_1} = \frac{t}{t-1}$. But it is easy to see (draw the graph!) that then $\frac{f_2}{f_1} = \frac{t'}{t'-1}$ has a single root in some interval $(x - \epsilon, x + \epsilon)$ for at least some t' close to t (slightly larger or slightly smaller). However, from Theorem 4.4.1 it follows that $f_{t'}$ has more than one root in the disc $|z - x| < \epsilon$, and so $f_{t'}$ would not have all roots real.

To end the proof, we apply the first step to f_ϵ and $f_{1-\epsilon}$ $(\epsilon > 0)$, which have only simple roots. Then we let $\epsilon \to 0$ and use Corollary 4.4.3 to obtain inequality (4.1). $\square$

4.5 Several variables: real stable polynomials

4.5.1 General facts

Denote $\mathbb{H} = \{z \in \mathbb{C} : \Im z > 0\}$.

Definition 4.5.1. *A polynomial $p(z_1, \ldots, z_m)$ is called* real stable *if it has real coefficients and it has no zeros in $\mathbb{H}^m$.*

In case $m = 1$ a real stable polynomial is a polynomial that has real coefficients and real zeros. Genuine examples in several variables are produced by the next lemma.

Lemma 4.5.2. *If $A_1, \ldots, A_m \in M_d(\mathbb{C})$ are positive matrices, then the polynomial*

$$q(z, z_1, \ldots, z_m) = \det(zI_d + \sum_{i=1}^{m} z_i A_i) \tag{4.1}$$

is real stable.

Proof. It is immediate from the definition that

$$q(\bar{z}, \bar{z}_1, \ldots, \bar{z}_m) = \overline{q(z, z_1, \ldots, z_m)},$$

whence the coefficients of q are real.

Assume that $q(z, z_1, \ldots, z_m) = 0$, and $\Im z, \Im z_i > 0$. Since $zI_d + \sum_{i=1}^{m} z_i A_i$ is not invertible, there exists $\xi \in \mathbb{C}^d$, $\xi \neq 0$, such that

$$0 = \langle (zI_d + \sum_{i=1}^{m} z_i A_i)\xi, \xi \rangle = z\|\xi\|^2 + \sum_{i=1}^{m} z_i \langle A_i \xi, \xi \rangle,$$

and so

$$0 = \Im z\|\xi\|^2 + \sum_{i=1}^{m} \Im z_i \langle A_i \xi, \xi \rangle.$$

This is a contradiction, since $\Im z\|\xi\|^2 > 0$ and $\Im z_i \langle A_i \xi, \xi \rangle \geq 0$ for all i. $\quad\square$

The next theorem gives the basic properties of real stable polynomials. Denote, for simplicity by ∂_i the partial derivative $\frac{\partial}{\partial z_i}$.

Theorem 4.5.3. *Suppose p is a real stable polynomial.*

(i) If $m > 1$ and $t \in \mathbb{R}$, then $p(z_1, \ldots, z_{m-1}, t)$ is either real stable or identically zero.

(ii) If $t \in \mathbb{R}$, then $(1 + t\partial_m)p$ is real stable.

130   *Recent Advances in Operator Theory and Operator Algebras*

Proof. (i) Obviously $p(z_1, \ldots, z_{m-1}, t)$ has real coefficients. Suppose it is not identically zero. If $\Im w > 0$ is fixed, then the polynomial $p(z_1, \ldots, z_{m-1}, w)$ is real stable by definition. Therefore all polynomials $p(z_1, \ldots, z_{m-1}, t + \frac{i}{n})$ (for $n \in \mathbb{N}$) are real stable. We let then $n \to \infty$ and apply Corollary 4.4.2 to $D = \mathbb{H}^m$ to obtain the desired result.

(ii) We may assume $t \neq 0$ (otherwise there is nothing to prove). Suppose $(1 + t\partial_m)p(z_1, \ldots, z_m) = 0$ for some $(z_1, \ldots, z_m) \in \mathbb{H}^m$. Since p is real stable, $p(z_1, \ldots, z_m) \neq 0$. The one-variable polynomial $q(z) := p(z_1, \ldots, z_{m-1}, z)$ has no roots with positive imaginary part (in particular, $q(z_m) \neq 0$), so we may write

$$q(z) = c \prod_{i=1}^{n}(z - w_i), \qquad \Im w_i \leq 0.$$

Therefore

$$0 = (1 + t\partial_m)p(z_1, \ldots, z_m) = (q + tq')(z_m) = q(z_m)\left(1 + t\frac{q'(z_m)}{q(z_m)}\right),$$

and, since $q(z_m) \neq 0$,

$$0 = 1 + t\sum_{i=1}^{n}\frac{1}{z_m - w_i} = 1 + t\sum_{i=1}^{n}\frac{\overline{z_m - w_i}}{|z_m - w_i|^2}.$$

Taking the imaginary part, we obtain

$$t\sum_{i=1}^{n}\frac{\Im w_i - \Im z_m}{|z_m - w_i|^2} = 0$$

which is a contradiction, since $t \neq 0$ and $\Im w_i - \Im z_m < 0$ for all i. $\square$

We will also need a lemma that uses a standard result in algebraic geometry, namely Bézout's theorem (which can be found in any standard text).

Lemma 4.5.4. *Suppose $p(z, w)$ is a nonconstant polynomial in two variables, of degree n in w, which is irreducible over $\mathbb{R}$. There is a finite set $F \in \mathbb{C}$ such that, if $p(z_0, w_0) = 0$ and $z_0 \notin F$, then*

1. the equation $p(z_0, w) = 0$ has n distinct solutions;

2. for each of these solutions (z_0, w_0) we have $\frac{\partial p}{\partial w}(z_0, w) \neq 0$.

Proof. First, if $p(z, w) = q(z)w^n + \ldots$, then the roots of q form a finite set F_1.

Second, if p is irreducible, then p and $\frac{\partial p}{\partial w}$ are coprime over $\mathbb{R}$, and hence also over $\mathbb{C}$. Bézout's theorem in algebraic geometry states that two curves defined by coprime equations have only a finite number of common points, so this is true about the sets defined by $p(z, w) = 0$ and $\frac{\partial p}{\partial w}(z, w) = 0$. Let F_2 be the set of the projections of these points onto the first coordinate. The set $F = F_1 \cup F_2$ has the required properties. $\square$

4.5.2 The barrier function

Our eventual purpose in this subsection is to obtain estimates on the roots of real stable polynomials; more precisely, we want to show that a restriction on the roots of a real stable polynomial p may imply a restriction on the roots of $(1 - \partial_i)p$ (which is also real stable by Theorem 4.5.3).

We will often use the restriction of a polynomial in m complex variables to $\mathbb{R}^m \subset \mathbb{C}^m$. To make things easier to follow, we will be consistent in this subsection with the following notation: z, w will belong to $\mathbb{C}^m$ (and corresponding subscripted letters in $\mathbb{C}$), while x, y, s, t will be in $\mathbb{R}^m$ (and corresponding subscripted letters in $\mathbb{R}$). If $x = (x_1, \ldots, x_m) \in \mathbb{R}^m$, then $\{y \geq x\}$ will denote $\{y = (y_1, \ldots, y_m) \in \mathbb{R}^m : y_i \geq x_i \text{ for all } i = 1, \ldots, m\}$.

The main tool is a certain function associated with p called the *barrier function*, whose one-dimensional version has already been met in Lemma 4.4.6. It is defined wherever $p \neq 0$ by $\Phi_p^i = \frac{\partial_i p}{p}$; if $p(x) > 0$ it can also be written as $\Phi_p^i(x) = \partial_i(\log p)(x)$. The argument of the barrier function will always actually be in $\mathbb{R}^m$.

The connection of the barrier function with our problem is given by the simple observation that if $p(x) \neq 0$ and $(1 - \partial_i)p(x) = 0$, then $\Phi_p^i(x) = 1$. So, in particular, a set on which $0 \leq \Phi_p^i < 1$ does not contain zeros of $(1 - \partial_i)p$. To determine such sets, the basic result is the next lemma, which is a multidimensional extension of Lemma 4.4.6.

Lemma 4.5.5. *Suppose $x \in \mathbb{R}^m$, and $p(z_1, \ldots, z_m)$ is a real stable polynomial that has no roots in $\{y \geq x\}$, then*

$$(-1)^k \frac{\partial^k}{\partial z_j^k} \Phi_p^i(x') \geq 0$$

for any $k \geq 0$, $1 \leq i, j \leq m$, and $x' \geq x$.

In particular, if e_j is one of the canonical basis vectors in $\mathbb{C}^m$, then $t \mapsto \Phi_p^i(x + te_j)$ is positive, nonincreasing, and convex on $[0, \infty]$.

Proof. The assertion reduces to Lemma 4.4.6 for $m = 1$ or for $k = 0$; and also for $i = j$, since then fixing all variables except the ith reduces the problem to the one variable case.

In the general case, it is enough to do it for $y = x$, since if p has no roots in $\{y \geq x\}$, then it has no roots in $\{y \geq x'\}$ for all $x' \geq x$. By fixing all variables except i and j, we may assume that $m = 2$, $i = 1$, $j = 2$, $k \geq 1$. Moreover, we may also assume that $p > 0$ on $\{y \geq x\}$ (otherwise we work with $-p$, since $\Phi_p^i = \Phi_{-p}^i$).

So we have to prove that, if $p(z_1, z_2)$ is a real stable polynomial which has

132   *Recent Advances in Operator Theory and Operator Algebras*

no zeros in $\{y_1 \geq x_1, \; y_2 \geq x_2\}$, then

$$0 \leq (-1)^k \frac{\partial^k}{\partial z_2^k} \Phi_p^1(x_1, x_2) = (-1)^k \frac{\partial^k}{\partial z_2^k} \left(\frac{\partial}{\partial z_1} \log p \right)(x_1, x_2)$$

$$= \frac{\partial}{\partial z_1} \left((-1)^k \frac{\partial^k}{\partial z_2^k} \log p \right)(x_1, x_2).$$

We will in fact prove that the map

$$t \mapsto (-1)^k \frac{\partial^k}{\partial z_2^k} \log p(t, x_2)$$

is increasing for $t \geq x_1$. It is enough to achieve this for p irreducible over $\mathbb{R}$, since, if $p = p_1 p_2$ is real stable and has no roots in $\{y \geq x\}$, then the same is true for p_1 and p_2, and obviously

$$(-1)^k \frac{\partial^k}{\partial z_2^k} \log p(t, x_2) = (-1)^k \frac{\partial^k}{\partial z_2^k} \log p_1(t, x_2) + (-1)^k \frac{\partial^k}{\partial z_2^k} \log p_2(t, x_2).$$

Suppose then that p is irreducible. For $t \geq x_1$ fixed, the polynomial $p(t, z)$ is real stable, and thus has all roots real; denote them, as in Section 4.4, by $\rho_1(t) \geq \cdots \geq \rho_n(t)$.

Applying to p Lemma 4.5.4, take $t \geq x_1$ that does not belong to the finite set F therein. The functions $\rho_i(t)$ are therefore differentiable in t, and we have

$$p(t, z) = c(t) \prod_{i=1}^{n} (z - \rho_i(t)). \tag{4.1}$$

Therefore

$$\left((-1)^k \frac{\partial^k}{\partial z_2^k} \log p \right) ((-1)^k \frac{\partial^k}{\partial z_2^k} \log p)(t, x_2) = (-1)^k \frac{\partial^k}{\partial z_2^k} \left(\sum_{i=1}^{n} \log(z - \rho_i(t)) \right) \Big|_{z=x_2}$$

$$= -\sum_{i=1}^{n} \frac{(k-1)!}{(x_2 - \rho_i(t))^k}. \tag{4.2}$$

If $t \geq x_1$, we cannot have $\rho_i(t) \geq x_2$ since then $(t, \rho_i(t))$ would be a root of p in $\{y \geq x\}$, contrary to the assumption. Thus $x_2 - \rho_i(t) > 0$, and in order to show that the function in (4.2) is increasing, it is enough to show that $t \mapsto \rho_i(t)$ is decreasing for $t \geq x_1$ and all i.

Now all ρ_is are differentiable for $t \geq x_1$, $t \notin F$. To show that they are decreasing, it is enough to show that $\rho_i'(t) \leq 0$ for such t. Suppose then that there exists $i \in \{1, \ldots, n\}$ and $t \geq x_1$ such that $\rho_i'(t) > 0$; let $s = \rho_i(t)$. Since $\frac{\partial p}{\partial z_2}(t, s) \neq 0$, we may apply the (complex) implicit function theorem in a neighborhood of (t, s) (in $\mathbb{C}^2$). We obtain that the solutions of $p(z_1, z_2) = 0$

The solution of the Kadison–Singer problem 133

therein are of the form $(z_1, g(z_1))$ for some locally defined analytic function of one variable g, which by analytic continuation has to be an extension of ρ to a complex neighborhood of t. So $g'(t) = \rho_i'(t)$, and in the neighborhood of t we have

$$g(z_1) = t + \rho_i'(t)(z_1 - t) + O(|z_1 - t|^2).$$

If $\Im z_1 > 0$ and small, one also has $\Im g(z_1) > 0$. We obtain thus the zero $(z_1, g(z_1))$ of p in $\mathbb{H}^2$, contradicting the real stability of p. This ends the proof of the lemma. $\qquad\square$

Corollary 4.5.6. *Suppose $x \in \mathbb{R}^m$, and p is a real stable polynomial, without zeros in $\{y \geq x\}$. Then $\Phi_p^j(y) \leq \Phi_p^j(x)$ for any $y \geq x$ and $j = 1, \ldots, m$.*

Proof. If p has no zeros in $\{y \geq x\}$, obviously it has no zeros in $\{y \geq x'\}$ for any $x' \geq x$. Therefore, by Lemma 4.5.5, the function $t \mapsto \Phi_p^j(x' + te_i)$ is nonincreasing on $[0, \infty)$ for any $i = 1, \ldots, m$. We have then

$$\Phi_p^j(x_1, \ldots, x_m) \geq \Phi_p^j(y_1, x_2, \ldots, x_m) \geq \Phi_p^j(y_1, y_2, x_3, \ldots, x_m)$$
$$\geq \cdots \geq \Phi_p^j(y_1, \ldots, y_m). \qquad\square$$

The main monotonicity and convexity properties of Φ_p^i are put to work in the next lemma to obtain a restriction on the location of zeros of $(1 - \partial_j)p$. As noted above, we will use the condition $\Phi_p^j < 1$, but in a more precise variant which will lend itself to iteration.

Lemma 4.5.7. *Let $x \in \mathbb{R}^m$, and p a real stable polynomial, without zeros in $\{y \geq x\}$. Suppose also that*

$$\Phi_p^j(x) + \frac{1}{\delta} \leq 1$$

for some $j \in \{1, \ldots, m\}$ and $\delta > 0$.
     Then

(i) $(1 - \partial_j)p$ has no zeros in $\{y \geq x\}$.

(ii) For any $i = 1, \ldots, m$ we have

$$\Phi_{(1-\partial_j)p}^i(x + \delta e_j) \leq \Phi_p^i(x).$$

Proof. By Corollary 4.5.6 we have

$$\frac{\partial_j p(y)}{p(y)} = \Phi(y) \leq \Phi(x) \leq 1 - \frac{1}{\delta} < 1,$$

so $\partial_j p(y) \neq p(y)$, or $(1 - \partial_j)p(y) \neq 0$.
     To prove (ii), note first that $(1 - \partial_j)p = p(1 - \Phi_p^j)$, whence $\log[(1 - \partial_j)p] = \log p + \log(1 - \Phi_p^j)$, so, by differentiating,

$$\Phi_{(1-\partial_j)p}^i = \Phi_p^i - \frac{\partial_i \Phi_p^j}{1 - \Phi_p^j}.$$

The required inequality becomes then

$$-\frac{\partial_i \Phi_p^j(x + \delta e_j)}{1 - \Phi_p^j(x + \delta e_j)} \le \Phi_p^i(x) - \Phi_p^i(x + \delta e_j). \tag{4.3}$$

By Corollary 4.5.6 we have

$$\Phi_p^j(x + \delta e_j) \le \Phi_p^j(x) \le 1 - \frac{1}{\delta},$$

or

$$\frac{1}{1 - \Phi_p^j(x + \delta e_j)} \le \delta.$$

Further on, p has no zeros in $\{y \ge x + \delta e_j\}$, so Lemma 4.5.5 (applied in $x + \delta e_j$) implies, in particular, that $-\partial_i \Phi_p^j(x + \delta e_j) \ge 0$, whence

$$-\frac{\partial_i \Phi_p^j(x + \delta e_j)}{1 - \Phi_p^j(x + \delta e_j)} \le -\delta \partial_i \Phi_p^j(x + \delta e_j).$$

To prove (4.3), it is then enough to show that

$$-\delta \partial_i \Phi_p^j(x + \delta e_j) \le \Phi_p^i(x) - \Phi_p^i(x + \delta e_j).$$

Using $\partial_i \Phi_p^j(x + \delta e_j) = \partial_j \Phi_p^i(x + \delta e_j)$, the inequality can be written

$$\Phi_p^i(x + \delta e_j) \le \Phi_p^i(x) + \delta \partial_j \Phi_p^i(x + \delta e_j).$$

This, however, is an immediate consequence of the convexity of the function $t \mapsto \Phi_p^i(x + te_j)$, that has been proved in Lemma 4.5.5. $\qquad\square$

Finally, the next theorem is the main result of this section that we will use in the sequel.

Theorem 4.5.8. *Let $x \in \mathbb{R}^m$, and p a real stable polynomial, without zeros in $\{y \ge x\}$. Suppose also that*

$$\Phi_p^j(x_1, \dots, x_m) + \frac{1}{\delta} \le 1$$

for some $\delta > 0$ and $j = 1, \dots, m$. Then

$$\prod_{i=1}^m (1 - \partial_i)p$$

has no zeros in $\{y \ge x + \tilde{\delta}\}$, where $\tilde{\delta} := (\delta, \dots, \delta) \in \mathbb{R}^m$.

Proof. The proof follows by applying Lemma 4.5.7 successively to $j = 1$ and x, then to $j = 2$ and $x + \delta e_1$, etc. $\qquad\square$

The solution of the Kadison–Singer problem   135

4.6 Characteristic and mixed characteristic polynomials

4.6.1 Mixed characteristic polynomial

We intend now to apply the results of Section 4.5 to polynomials related to matrices. Our final goal is to estimate eigenvalues; that is, roots of the characteristic polynomial. But we will first consider another polynomial, attached to a tuple of matrices.

Definition 4.6.1. *If $A_1, \ldots, A_m \in M_d(\mathbb{C})$, then the* mixed characteristic *polynomial of the matrices A_i is defined by the formula*

$$\mu[A_1, \ldots, A_m](z) = \prod_{i=1}^{m}(1 - \partial_i)\det\left(zI_d + \sum_{i=1}^{m} z_i A_i\right)\Bigg|_{z_1 = \cdots = z_m = 0}. \qquad (4.1)$$

It is easily seen that if we fix $m - 1$ of the matrices $A_1, \ldots, A_m$, then $\mu[A_1, \ldots, A_m](z)$ is of degree 1 in the entries of the remaining matrix. Indeed, if we develop the determinant that enters (4.1), then any term that contains a product of, say, k entries of A_j also contains the factor z_j^k. If we apply $(1 - \partial_j)$, we are left with z_j^{k-1}, and if $k \geq 2$ this terms becomes 0 if $z_j = 0$.

Example 4.6.2. *For one or two matrices we have*

$$\mu[A_1](z) = z^d - z^{d-1}\mathrm{tr}A_1 \quad \text{if } m = 1,$$

$$\mu[A_1, A_2](z) = z^d - z^{d-1}(\mathrm{tr}A_1 + \mathrm{tr}A_2) + z^{d-2}(\mathrm{tr}A_1\mathrm{tr}A_2 - \mathrm{tr}(A_1 A_2))$$

if $m = 2$. In the general case, the coefficients of $\mu[A_1, \ldots, A_m](z)$ are certain expressions in the traces of monomials in $A_1, \ldots, A_m$ that are well known in the invariant theory of matrices (see [13]).

The results in Section 4.5 have consequences for the mixed characteristic polynomials.

Theorem 4.6.3. *Suppose $A_1, \ldots, A_m \in M_d(\mathbb{C})$ are positive matrices. Then $\mu[A_1, \ldots, A_m](z)$ has only real roots.*

Proof. We have seen in Lemma 4.5.2 that the polynomial q defined by (4.1) is real stable. But $\mu[A_1, \ldots, A_m]$ is obtained from q by first applying $(1 - \partial_i)$ for $i = 1, \ldots, m$ and then specializing to $z_1 = \cdots = z_m = 0$. By Theorem 4.5.3, these operations preserve the real stable character. So $\mu[A_1, \ldots, A_m]$ is a real stable polynomial of one variable, which means exactly that it has real roots. $\square$

Remember Jacobi's formula for the derivative of the determinant of an invertible matrix:

$$\frac{(\det M(t))'}{\det M(t)} = \mathrm{tr}\left(M(t)^{-1}M'(t)\right). \qquad (4.2)$$

Theorem 4.6.4. *Suppose* $A_1, \ldots, A_m \in M_d(\mathbb{C})$ *are positive matrices, such that* $\sum_{i=1}^m A_i = I_d$ *and* $\mathrm{tr} A_i \leq \epsilon$ *for each* $i = 1, \ldots, m$. *Then any root of* $\mu[A_1, \ldots, A_m]$ *is smaller than* $(1 + \sqrt{\epsilon})^2$.

Proof. The polynomial

$$p(z) := \det(\sum_{i=1}^m z_i A_i)$$

is real stable, being the specialization of the polynomial q in (4.1) to $z = 0$. If $t > 0$ and $\tilde{t} := (t, \ldots, t) \in \mathbb{C}^d$, then, for $y \geq \tilde{t}$ we have $\sum_{i=1}^m y_i A_i \geq \sum_{i=1}^m t A_i = t I_d$. Therefore $\sum_{i=1}^m y_i A_i$ is invertible, and $p(y) \neq 0$.

We may apply Jacobi's formula (4.2) in order to compute the barrier function Φ_p^j, and we obtain

$$\Phi_p^j(x_1, \ldots, x_m) = \mathrm{tr}((\sum_{i=1}^m z_i A_i)^{-1} A_j).$$

In particular, if $t > 0$, then

$$\Phi_p^j(t, \ldots, t) = \mathrm{tr}(t^{-1} A_j) \leq \frac{\epsilon}{t}.$$

It follows then from Theorem 4.5.8 that if $t, \delta > 0$ are such that $\frac{\epsilon}{t} + \frac{1}{\delta} \leq 1$, then $\prod_{i=1}^m (1 - \partial_i) p$ has no zeros in $\{y \geq (t + \delta, \ldots, t + \delta)\}$. The choice $t = \epsilon + \sqrt{\epsilon}$, $\delta = 1 + \sqrt{\epsilon}$ (which can easily be shown to be optimal) yields $t + \delta = (1 + \sqrt{\epsilon})^2$, and therefore p has no roots y with $y_i \geq (1 + \sqrt{\epsilon})^2\}$ for all i.

Now, using the relation $\sum_{i=1}^m A_i = 1$, one obtains

$$\mu[A_1, \ldots, A_m](z) = \prod_{i=1}^m (1 - \partial_i) \det(z I_d + \sum_{i=1}^m z_i A_i)\big|_{z_1 = \cdots = z_m = 0}$$

$$= \prod_{i=1}^m (1 - \partial_i) \det(\sum_{i=1}^m w_i A_i)\big|_{w_1 = \cdots = w_m = z}$$

$$= \prod_{i=1}^m (1 - \partial_i) p(z, z, \ldots, z),$$

which cannot be zero if $z \geq (1 + \sqrt{\epsilon})^2$. Therefore all roots of μ are smaller than $(1 + \sqrt{\epsilon})^2$. $\square$

4.6.2 Decomposing in rank one matrices and the characteristic polynomial

In an important particular case the mixed characteristic polynomial coincides with a usual characteristic polynomial. Remember this is defined, for $A \in M_d(\mathbb{C})$, by $p_A(z) = \det(z I_d - A)$.

The solution of the Kadison–Singer problem 137

Lemma 4.6.5. *Suppose $B, A_1, \ldots, A_m \in M_d(\mathbb{C})$, and $A_1, \ldots, A_m$ have rank one. Then the polynomial*

$$(z_1, \ldots, z_m) \mapsto \det(B + z_1 A_1 + \cdots + z_m A_m)$$

is of degree ≤ 1 separately in each variable.

Proof. By fixing all the variables except one, we have to show that, for any $B, A_1 \in M_d(\mathbb{C})$, A_1 of rank one, the function

$$z \mapsto \det(B + z A_1)$$

is of degree at most 1. This is obvious if we choose a basis in which the first vector spans the image of A_1, and we develop the determinant with respect to the first row. $\qquad\square$

Suppose now $p(z_1, \ldots, z_m)$ is a polynomial of degree ≤ 1 separately in each variable. Then p is equal to its Taylor expansion at the origin of order 1 in each variable, that is,

$$p(z_1, \ldots, z_m) = \sum_{\epsilon_i \in \{0,1\}} c_{\epsilon_1, \ldots, \epsilon_m} z_1^{\epsilon_1} \cdots z_m^{\epsilon_m},$$

with

$$c_{\epsilon_1, \ldots, \epsilon_m} = \partial_1^{\epsilon_1} \cdots \partial_m^{\epsilon_m} p(w_1, \ldots, w_m)\big|_{w_1 = \cdots = w_m = 0}.$$

Therefore

$$p(z_1, \ldots, z_m) = \sum_{\epsilon_i \in \{0,1\}} z_1^{\epsilon_1} \cdots z_m^{\epsilon_m} \partial_1^{\epsilon_1} \cdots \partial_m^{\epsilon_m} p(w_1, \ldots, w_m)\big|_{w_1 = \cdots = w_m = 0}$$

$$= \prod_{i=1}^{m} (1 + z_i \partial_i) p(w_1, \ldots, w_m)\big|_{w_1 = \cdots = w_m = 0}.$$

In the case of the polynomial in Lemma 4.6.5, this formula becomes

$$\det\Big(B + \sum_{i=1}^{m} z_i A_i\Big) = \prod_{i=1}^{m} (1 + z_i \partial_i) \det\Big(B + \sum_{i=1}^{m} w_i A_i\Big)\Big|_{w_1 = \cdots = w_m = 0}.$$

In fact, we are interested by this last formula precisely when $B = z I_d$ and all $z_i = -1$. We obtain then the next theorem.

Theorem 4.6.6. *Suppose $A_1, \ldots, A_m \in M_d(\mathbb{C})$ have rank one. If $A = A_1 + \cdots + A_m$, then*

$$p_A(z) = \mu[A_1, \ldots, A_m](z).$$

Remark 4.6.7. *The mixed characteristic polynomial and the usual characteristic polynomial are invariant with respect to a change of basis. So, although we have spoken about matrices for convenience, the statements of Theorems 4.6.4 and 4.6.6 can be stated for $A_1, \ldots, A_m \in \mathcal{L}(V)$, where $\mathcal{L}(V)$ denotes the space of linear operators on the finite-dimensional vector space V.*

4.7 Randomization

4.7.1 Random matrices and determinants

Let $(\Omega, \mathbf{p})$ be a finite probability space. If X is a random variable on Ω, the *expectation* (or *average*) of $\mathbb{E}(X)$ is defined, usually, by

$$\mathbb{E}(X) := \sum_{\omega \in \Omega} \mathbf{p}(\omega) X(\omega).$$

If $X_1, \ldots, X_m$ are *independent* random variables, then, in particular, we have

$$\mathbb{E}(X_1 \cdots X_m) = \mathbb{E}(X_1) \cdots \mathbb{E}(X_m). \tag{4.1}$$

We will use random matrices $\mathbf{A}(\omega) \in M_d(\mathbb{C})$, whose entries are random variables; then $\mathbb{E}(\mathbf{A})$ is the matrix whose entries are the expectations of the corresponding entries of $\mathbf{A}$. The random matrices $\mathbf{A}_1, \mathbf{A}_2$ are called independent if any entry of $\mathbf{A}_1$ is independent of every entry of $\mathbf{A}_2$. Also, when we say that a random matrix $\mathbf{A}(\omega)$ has rank one, this means that $\mathbf{A}(\omega)$ has rank one for any $\omega \in \Omega$.

The characteristic polynomial $p_{\mathbf{A}}$ of a random matrix $\mathbf{A}$ is also a random variable, by which we mean that its coefficients are random variables. Then the polynomial $\mathbb{E}(p_{\mathbf{A}})$ has as coefficients the expectations of the coefficients of $\mathbf{A}$.

Theorem 4.7.1. *Suppose* $\mathbf{A}_1(\omega), \ldots, \mathbf{A}_m(\omega)$ *are independent rank one random matrices in* $M_d(\mathbb{C})$, *and* $\mathbf{A} = \mathbf{A}_1 + \cdots + \mathbf{A}_m$. *Then*

$$\mathbb{E}(p_{\mathbf{A}}) = \mu[\mathbb{E}(\mathbf{A}_1), \ldots, \mathbb{E}(\mathbf{A}_m)].$$

Proof. By Theorem 4.6.6 we have, for each $\omega \in \Omega$, $p_{\mathbf{A}(\omega)} = \mu[\mathbf{A}_1(\omega), \ldots, \mathbf{A}_m(\omega)]$. By taking expectations,

$$\mathbb{E}(p_{\mathbf{A}}) = \mathbb{E}(\mu[\mathbf{A}_1(\omega), \ldots, \mathbf{A}_m(\omega)]).$$

Independence of $\mathbf{A}_i$s combined with the fact that μ has degree at most 1 in the entries of the matrices gives

$$\mathbb{E}(\mu[\mathbf{A}_1(\omega), \ldots, \mathbf{A}_m(\omega)]) = \mu[\mathbb{E}(\mathbf{A}_1), \ldots, \mathbb{E}(\mathbf{A}_m)],$$

which ends the proof. $\square$

We can say more if we also assume that the $\mathbf{A}_i$s are all positive.

Theorem 4.7.2. *Suppose* $\mathbf{A}_1(\omega), \ldots, \mathbf{A}_m(\omega)$ *are independent rank one positive random matrices in* $M_d(\mathbb{C})$, *and* $\mathbf{A} = \mathbf{A}_1 + \cdots + \mathbf{A}_m$. *Then, for any* $j = 1, \ldots, d$, *we have*

$$\min_{\omega \in \Omega} \rho_j(p_{\mathbf{A}(\omega)}) \leq \rho_j(\mu[\mathbb{E}(\mathbf{A}_1), \ldots, \mathbb{E}(\mathbf{A}_m)]) \leq \max_{\omega \in \Omega} \rho_j(p_{\mathbf{A}(\omega)}).$$

Proof. We prove only the left-hand side inequality; the right-hand side is similar. It is enough to show that for any $i = 1, \ldots$ we have

$$\min_{\omega \in \Omega} \rho_j(\mu[\mathbf{A}_1(\omega), \ldots, \mathbf{A}_{i-1}(\omega), \mathbf{A}_i(\omega), \mathbb{E}(\mathbf{A}_{i+1}), \ldots, \mathbb{E}(\mathbf{A}_m)])$$

$$\leq \min_{\omega \in \Omega} \rho_j(\mu[\mathbf{A}_1(\omega), \ldots, \mathbf{A}_{i-1}(\omega), \mathbb{E}(\mathbf{A}_i), \mathbb{E}(\mathbf{A}_{i+1}), \ldots, \mathbb{E}(\mathbf{A}_m)]).$$

$$(4.2)$$

Indeed, for $i = m$ the left-hand side coincides with $\min_{\omega \in \Omega} \rho_j(p_{\mathbf{A}(\omega)})$ by Theorem 4.6.6, while for $i = 1$ the right-hand side is precisely $\rho_j(\mu[\mathbb{E}(\mathbf{A}_1), \ldots, \mathbb{E}(\mathbf{A}_m)])$. The chain of inequalities corresponding to $i = 1, 2, \ldots, m$ then proves the theorem.

Fix then i and $\omega \in \Omega$, and consider the family of polynomials

$$f_{\omega'} = \mu[\mathbf{A}_1(\omega), \ldots, \mathbf{A}_{i-1}(\omega), \mathbf{A}_i(\omega'), \mathbb{E}(\mathbf{A}_{i+1}), \ldots, \mathbb{E}(\mathbf{A}_m)], \quad \omega' \in \Omega.$$

Take $c_{\omega'} \geq 0$, with $\sum_{\omega' \in \Omega} c_{\omega'} = 1$. Using the fact that the mixed characteristic polynomial has degree at most 1 in the entries of the matrices, we obtain that

$$\sum_{\omega' \in \Omega} c_{\omega'} f_{\omega'} = \sum_{\omega' \in \Omega} c_{\omega'} \mu[\mathbf{A}_1(\omega), \ldots, \mathbf{A}_{i-1}(\omega), \mathbf{A}_i(\omega'), \mathbb{E}(\mathbf{A}_{i+1}), \ldots, \mathbb{E}(\mathbf{A}_m)]$$

$$= \mu[\mathbf{A}_1(\omega), \ldots, \mathbf{A}_{i-1}(\omega), \sum_{\omega' \in \Omega} c_{\omega'} \mathbf{A}_i(\omega'), \mathbb{E}(\mathbf{A}_{i+1}), \ldots, \mathbb{E}(\mathbf{A}_m)].$$

Since the last polynomial is the mixed characteristic polynomial of positive matrices, it has all roots real by Theorem 4.6.3. It follows by Theorem 4.4.9 that $\{f_{\omega'} : \omega' \in \Omega\}$ is a nice family. Moreover, if we take as coefficients of the convex combination $c_{\omega'} = \mathbf{p}(\omega')$ and so for any $j = 1, \ldots, d$ we have $\sum_{\omega' \in \Omega} c_{\omega'} \mathbf{A}_i(\omega') = \mathbb{E}(\mathbf{A}_i)$. Applying the last part of Theorem 4.4.9 it follows that for any $j = 1, \ldots, d$,

$$\min_{\omega' \in \Omega} \rho_j(\mu[\mathbf{A}_1(\omega), \ldots, \mathbf{A}_{i-1}(\omega), \mathbf{A}_i(\omega'), \mathbb{E}(\mathbf{A}_{i+1}), \ldots, \mathbb{E}(\mathbf{A}_m)])$$

$$\rho_j(\mu[\mathbf{A}_1(\omega), \ldots, \mathbf{A}_{i-1}(\omega), \mathbb{E}(\mathbf{A}_i), \mathbb{E}(\mathbf{A}_{i+1}), \ldots, \mathbb{E}(\mathbf{A}_m)]).$$

Taking the minimum with respect to $\omega \in \Omega$, we obtain

$$\min_{\omega \in \Omega} \min_{\omega' \in \Omega} \rho_j(\mu[\mathbf{A}_1(\omega), \ldots, \mathbf{A}_{i-1}(\omega), \mathbf{A}_i(\omega'), \mathbb{E}(\mathbf{A}_{i+1}), \ldots, \mathbb{E}(\mathbf{A}_m)])$$

$$\min_{\omega \in \Omega} \rho_j(\mu[\mathbf{A}_1(\omega), \ldots, \mathbf{A}_{i-1}(\omega), \mathbb{E}(\mathbf{A}_i), \mathbb{E}(\mathbf{A}_{i+1}), \ldots, \mathbb{E}(\mathbf{A}_m)]).$$

$$(4.3)$$

Suppose the minimum in the left-hand side is attained in $\omega = \omega_0$, $\omega' = \omega_0'$. By independence of the random matrices $\mathbf{A}_i$, we have

$$\mathbf{p}(\{\sigma \in \Omega : \mathbf{A}_1(\sigma) = \mathbf{A}_1(\omega_0), \ldots, \mathbf{A}_{i-1}(\sigma) = \mathbf{A}_{i-1}(\omega_0), \mathbf{A}_i(\sigma) = \mathbf{A}_{(\omega_0')}\})$$

$$\mathbf{p}(\{\sigma \in \Omega : \mathbf{A}_1(\sigma) = \mathbf{A}_1(\omega_0), \ldots, \mathbf{A}_{i-1}(\sigma) = \mathbf{A}_{i-1}(\omega_0)\})$$

$$\mathbf{p}(\{\sigma \in \Omega : \mathbf{A}_i(\sigma) = \mathbf{A}_{(\omega_0')}\}) > 0.$$

140   *Recent Advances in Operator Theory and Operator Algebras*

Taking $\sigma_0 \in \Omega$ in the set in the left-hand side, we obtain

$$\min_{\sigma \in \Omega} \rho_j(\mu[\mathbf{A}_1(\sigma), \dots, \mathbf{A}_{i-1}(\sigma), \mathbf{A}_i(\sigma), \mathbb{E}(\mathbf{A}_{i+1}), \dots, \mathbb{E}(\mathbf{A}_m)])$$

$$\leq \rho_j(\mu[\mathbf{A}_1(\sigma_0), \dots, \mathbf{A}_{i-1}(\sigma_0), \mathbf{A}_i(\sigma_0), \mathbb{E}(\mathbf{A}_{i+1}), \dots, \mathbb{E}(\mathbf{A}_m)])$$

$$= \rho_j(\mu[\mathbf{A}_1(\omega_0), \dots, \mathbf{A}_{i-1}(\omega_0), \mathbf{A}_i(\omega_0'), \mathbb{E}(\mathbf{A}_{i+1}), \dots, \mathbb{E}(\mathbf{A}_m)])$$

$$= \min_{\omega \in \Omega} \min_{\omega' \in \Omega} \rho_j(\mu[\mathbf{A}_1(\omega), \dots, \mathbf{A}_{i-1}(\omega), \mathbf{A}_i(\omega'), \mathbb{E}(\mathbf{A}_{i+1}), \dots, \mathbb{E}(\mathbf{A}_m)]).$$

This inequality, together with (4.3), implies (4.2), finishing thus the proof of the theorem. $\square$

Remark 4.7.3. *The point of Theorem 4.7.2 is that the middle term might be easier to compute or to estimate. But, since the matrices $\mathbb{E}(\mathbf{A}_1), \dots, \mathbb{E}(\mathbf{A}_m)$ are* not *of rank one, Theorem 4.6.6 does not apply, and $\mu[\mathbb{E}(\mathbf{A}_1), \dots, \mathbb{E}(\mathbf{A}_m)]$ is not a characteristic polynomial. However, Theorem 4.7.2 tells us that its roots can be used to estimate the eigenvalues of $\mathbf{A}(\omega)$ for at least some value of ω.*

Corollary 4.7.4. *Let $\mathbf{A}_1(\omega), \dots, \mathbf{A}_m(\omega)$ be independent rank one positive random matrices in $M_d(\mathbb{C})$, and $\mathbf{A} = \mathbf{A}_1 + \cdots + \mathbf{A}_m$. Suppose $\mathbb{E}(\mathbf{A}) = I_d$ and $\mathbb{E}(\mathrm{tr}\mathbf{A}_i) \leq \epsilon$ for some $\epsilon > 0$. Then*

$$\min_{\omega \in \Omega} \|\mathbf{A}(\omega)\| \leq (1 + \sqrt{\epsilon})^2.$$

Proof. Since $\mathrm{tr}(\mathbb{E}(\mathbf{A}_i)) = \mathbb{E}(\mathrm{tr}\mathbf{A}_i) \leq \epsilon$, the matrices $\mathbb{E}(\mathbf{A}_1), \dots, \mathbb{E}(\mathbf{A}_m)$ satisfy the hypotheses of Theorem 4.6.4, and all roots of $\mu[\mathbb{E}(\mathbf{A}_1), \dots, \mathbb{E}(\mathbf{A}_m)]$ are smaller than $(1 + \sqrt{\epsilon})^2$. By Theorem 4.7.2 we obtain, in particular,

$$\min_{\omega \in \Omega} \|\mathbf{A}(\omega)\| = \min_{\omega \in \Omega} \rho_1(p_{\mathbf{A}(\omega)}) \leq \rho_1(\mu[\mathbb{E}(\mathbf{A}_1), \dots, \mathbb{E}(\mathbf{A}_m)]) \leq (1 + \sqrt{\epsilon})^2. \quad \square$$

4.7.2   Probability and partitions

The last theorem of this section gets us closer to the paving conjecture. It is here that we make the connection between the probability space and the partitions. Let us first note that, similarly to Remark 4.6.7, one can see that the independence condition is not affected by a change of basis. So in Theorem 4.7.2 and in Corollary 4.7.4 we may assume that $\mathbf{A}_i$ take values in $B(V)$ for some finite-dimensional vector space V. This observation will be used in the proof of the next theorem.

Theorem 4.7.5. *Suppose $A_1, \dots, A_m \in M_d(\mathbb{C})$ are positive rank one matrices, such that $\sum_{i=1}^m A_i = I_d$ and $\|A_i\| \leq C$ for all $i = 1, \dots, m$. Then for every positive integer r there exists a partition $S_1, \dots, S_r$ of $\{1, \dots, m\}$, such that*

$$\left\| \sum_{i \in S_j} A_i \right\| \leq \left(\frac{1}{\sqrt{r}} + \sqrt{C} \right)^2$$

for any $j = 1, \dots, r$.

The solution of the Kadison–Singer problem 141

Proof. Since the purpose is to find a partition with certain properties, we will take as a random space Ω precisely the space of all partitions of $\{1,\dots,m\}$ in r sets, with uniform probability $\mathbf{p}$. Such a partition is determined by an element $\omega = (\omega_1,\dots,\omega_m)$, where $\omega_j \in \{1,\dots,r\}$ and $S_j = \{k : \omega_k = j\}$; so $\Omega = \{1,\dots,r\}^m$. Also, the different coordinates, that is, the maps $\omega \mapsto \omega_i$, are independent scalar random variables on Ω.

We consider the space $V := \mathbb{C}^d \oplus \cdots \oplus \mathbb{C}^d$ (r summands) and define the random matrices $\mathbf{A}_i$ ($i = 1,\dots,m$) with values in $\mathcal{L}(V)$ by

$$\mathbf{A}_i(\omega) = 0 \oplus \cdots \oplus rA_i \oplus \cdots \oplus 0, \tag{4.1}$$

where rA_i appears in position ω_i.

These are independent random matrices (since the coordinates ω_i are independent). If we fix $1 \le j \le r$, then $\omega_i = j$ with probability $1/r$, and so rA_i appears in position j with probability $1/r$. Therefore

$$\mathbb{E}(\mathbf{A}_i) = \frac{1}{r}rA_i \oplus \cdots \oplus \frac{1}{r}rA_i = A_i \oplus A_i \oplus \cdots \oplus A_i.$$

If $\mathbf{A} = \mathbf{A}_1 + \cdots + \mathbf{A}_m$, then

$$\mathbb{E}(\mathbf{A}) = \sum_{i=1}^{m} \mathbb{E}(\mathbf{A}_i) = \sum_{i=1}^{m} (A_i \oplus A_i \oplus \cdots \oplus A_i) = I_V.$$

Since $\mathrm{tr}\mathbf{A}_i(\omega) = r\mathrm{tr}A_i$ for all i, we have

$$\mathbb{E}(\mathrm{tr}\mathbf{A}_i) = \mathbb{E}(r\mathrm{tr}A_i) = r\mathbb{E}(\|A_i\|) \le rC.$$

Corollary 4.7.4 yields the existence of $\omega \in \Omega$ such that

$$\|\mathbf{A}(\omega)\| \le (1 + \sqrt{rC})^2.$$

But, according to (4.1), we have

$$\mathbf{A}(\omega) = \left(r\sum_{\omega_i=1} A_i\right) \oplus \left(r\sum_{\omega_i=2} A_i\right) \oplus \cdots \oplus \left(r\sum_{\omega_i=r} A_i\right).$$

We define then $S_j = \{i : \omega_i = j\}$. It follows that $\|r\sum_{i\in S_j} A_i\| \le (1 + \sqrt{rC})^2$ for all j, and dividing by r ends the proof of the theorem. $\qquad\square$

4.8 Proof of the paving conjecture

We may now proceed to the proof of the paving conjecture; from this point on all we need from the previous sections is Theorem 4.7.5. We first

142   *Recent Advances in Operator Theory and Operator Algebras*

deal with orthogonal projections. For such operators the paving conjecture is trivially verified (exercise: if P is an orthogonal projection and $\operatorname{diag} P = 0$, then $P = 0$). But we will prove a quantitative version of the paving conjecture, in which one does not assume zero diagonal.

Lemma 4.8.1. *Suppose $P \in M_m(\mathbb{C})$ is an orthogonal projection. For any $r \in \mathbb{N}$ there exist diagonal orthogonal projections $Q_1, \ldots, Q_r \in M_m(\mathbb{C})$, with $\sum_{j=1}^r Q_j = I_m$, such that*

$$\|Q_j P Q_j\| \le \left(\frac{1}{\sqrt{r}} + \sqrt{\|\operatorname{diag} P\|} \right)^2$$

for all $j = 1, \ldots, r$.

Proof. Denote by V the image of P, and $d = \dim V$. Let $(e_i)_{i=1}^m$ be a basis in $\mathbb{C}^m$, and define on V the rank one positive operators $A_i = P(e_i) \otimes P(e_i)$ (that is, $A_i(v) = \langle v, P(e_i) \rangle P(e_i)$). We have

$$\|A_i\| = \|P(e_i)\|^2 = \langle P(e_i), e_i \rangle \le \|\operatorname{diag} P\|, \tag{4.1}$$

and

$$\sum_{i=1}^m A_i = \sum_{i=1}^m P(e_i) \otimes P(e_i) = P(\sum_{i=1}^m e_i \otimes e_i)P = I_V. \tag{4.2}$$

From (4.1) and (4.2) it follows that A_i satisfy the hypotheses of Theorem 4.7.5, with $C = \|\operatorname{diag} P\|_\infty$. There exists therefore a partition $S_1, \ldots, S_r$ of $\{1, \ldots, m\}$, such that

$$\left\| \sum_{i \in S_j} A_i \right\| \le \left(\frac{1}{\sqrt{r}} + \sqrt{\|\operatorname{diag} P\|_\infty} \right)^2$$

for any $j = 1, \ldots, r$.

Define then $Q_j \in M_m(\mathbb{C})$ to be the diagonal orthogonal projection on the span of $\{e_i : i \in S_j\}$, that is, $Q_j = \sum_{i \in S_j} e_i \otimes e_i$. Therefore

$$P Q_j P = \sum_{i \in S_j} P(e_i) \otimes P(e_i) = \sum_{i \in S_j} A_i.$$

Then

$$\|Q_j P Q_j\| = \|Q_j P (Q_j P)^*\| = \|(Q_j P)^* Q_j P\|^2$$

$$= \|P Q_j P\| = \left\| \sum_{i \in S_j} A_i \right\| \le \left(\frac{1}{\sqrt{r}} + \sqrt{\|\operatorname{diag} P\|_\infty} \right)^2.$$

and the lemma is proved. $\qquad\qquad\square$

The solution of the Kadison–Singer problem 143

Theorem 4.8.2 (The paving conjecture). *For any $\epsilon > 0$ there exists $r \in \mathbb{N}$ such that, for any $m \in \mathbb{N}$ and $T \in M_m(\mathbb{C})$ with $\operatorname{diag} T = 0$, there exist diagonal orthogonal projections $Q_1, \ldots, Q_r \in M_m(\mathbb{C})$, with $\sum_{j=1}^{r} Q_j = I_m$, such that*

$$\|Q_j T Q_j\| \le \epsilon \|T\|$$

for all $j = 1, \ldots, r$.

Proof. Suppose first that $T = T^*$ and $\|T\| \le 1$. The $2m \times 2m$ matrix

$$P = \begin{pmatrix} \frac{I_m + T}{2} & \frac{1}{2}(I_m - T^2)^{1/2} \\ \frac{1}{2}(I_m - T^2)^{1/2} & \frac{I_m - T}{2} \end{pmatrix}$$

is an orthogonal projection and $\operatorname{diag} P = (\frac{1}{2}, \ldots, \frac{1}{2})$. Choose r large enough to have $2 \left(\frac{1}{\sqrt{r}} + \frac{1}{\sqrt{2}} \right)^2 - 1 \le \epsilon$. It follows from Lemma 4.8.1 that there exist diagonal projections $Q_1'', \ldots, Q_r'' \in M_{2m}(\mathbb{C})$ with $\sum_{i=1}^{r} Q_i'' = I_{2d}$ and $\|Q_i'' P Q_i''\| \le \left(\frac{1}{\sqrt{r}} + \frac{1}{\sqrt{2}} \right)^2$ for all $i = 1, \ldots, r$.

Let $Q_i'' = Q_i + Q_i'$ be the decomposition of Q_i'' in the diagonal projections corresponding to the first m and the last m vectors of the basis of $\mathbb{C}^{2m}$. So $\sum_{i=1}^{r} Q_i = \sum_{i=1}^{r} Q_i = I_m$ and, for each $i = 1, \ldots, m$,

$$\|Q_i(I + T)Q_i\| \le 2 \left(\frac{1}{\sqrt{r}} + \frac{1}{\sqrt{2}} \right)^2, \qquad \|Q_i'(I - T)Q_i'\| \le 2 \left(\frac{1}{\sqrt{r}} + \frac{1}{\sqrt{2}} \right)^2. \tag{4.3}$$

The first inequality implies that $Q_i(I + T)Q_i \le 2 \left(\frac{1}{\sqrt{r}} + \frac{1}{\sqrt{2}} \right)^2 Q_i$, so

$$-Q_i \le Q_i T Q_i \le \left[2 \left(\frac{1}{\sqrt{r}} + \frac{1}{\sqrt{2}} \right)^2 - 1 \right] Q_i \le \epsilon Q_i \tag{4.4}$$

(the left inequality being obvious). Similarly, the second inequality in (4.3) yields

$$-\epsilon Q_i' \le \left[1 - 2 \left(\frac{1}{\sqrt{r}} + \frac{1}{\sqrt{2}} \right)^2 \right] Q_i' \le Q_i' T Q_i' \le Q_i'. \tag{4.5}$$

If we define $Q_{ij} = Q_i Q_j'$ $(i, j = 1, \ldots, r)$, then $\sum_{i,j=1}^{r} Q_{ij} = I_m$, and it follows from (4.4) and (4.5) that

$$-\epsilon Q_{ij} \le Q_{ij} T Q_{ij} \le \epsilon Q_{ij},$$

or $Q_{ij} T Q_{ij} \le \epsilon$. The theorem is thus proved for T, a selfadjoint contraction, and it is immediate to extend it to arbitrary selfadjoint matrices.

If we take now an arbitrary $T \in M_m(\mathbb{C})$, with $\operatorname{diag} T = 0$, we may write it as $T = A + iB$, with A, B selfadjoint, $\|A\|, \|B\| \le \|T\|$, and $\operatorname{diag} A = \operatorname{diag} B = 0$. Applying the first step, one finds diagonal projections $Q_1', \ldots, Q_r', Q_1'', \ldots, Q_r'' \in M_m(\mathbb{C})$, with $\sum_{i=1}^{r} Q_i' = \sum_{i=1}^{r} Q_i'' = I_m$,

144   *Recent Advances in Operator Theory and Operator Algebras*

$\|Q_i'AQ_i'\| \leq \frac{\epsilon}{2}\|T\|$, and $\|Q_i''BQ_i''\| \leq \frac{\epsilon}{2}\|T\|$ for $i = 1, \ldots, r$. If we define $Q_{ij} = Q_i'Q_j''$, then $\sum_{i,j=1}^r Q_{i,j} = I_m$, and $\|Q_{ij}TQ_{ij}\| \leq \epsilon\|T\|$ for $i, j = 1, \ldots, r$. $\square$

By writing carefully the estimates in the proof, one sees also that we may take r of order ϵ^{-4}.

4.9   Final remarks

1. As noted in Remark 4.2.6, there is a connection between the Kadison–Singer problem and quantum mechanics. We will give here a very perfunctory account. In the von Neumann picture of quantum mechanics, states (in the common sense) of a system correspond to states φ (in the C^*-algebra sense) of $B(\mathcal{H})$, while observables of the system correspond to selfadjoint operators $A \in B(\mathcal{H})$. The value of an observable in a state is precisely $\varphi(A)$.

A maximal abelian C^*-algebra $\mathcal{A} \subset B(\mathcal{H})$ corresponds to a maximal set of mutually compatible observables. If the extension of any pure state on $\mathcal{A}$ to a state on $B(\mathcal{H})$ is unique, then one can say that the given set of observables determines completely all other observables. This seems to have been assumed by Dirac implicitly.

Now, there are various maximal abelian subalgebras of $B(\mathcal{H})$, but the problem can essentially be reduced to two different basic types: continuous (that are essentially isomorphic to L^∞ acting as multiplication operators on L^2) and discrete (that are isomorphic to $\mathcal{D}$ acting in ℓ^2). The main topic of the original paper [7] is to prove that extension of pure states is *not* unique in general for continuous subalgebras. They suspected that the same thing happens for the discrete case, but could not prove it, and so posed it as an open problem.

2. We have said in the introduction that there are many statements that had been shown to be equivalent to (KS), besides (PC) that we have used in an essential way. We have thus, among others,

1. Weaver's conjectures in discrepancy theory. The original proof in [10] goes actually through one of these; the shortcut using (PC) is due to Tao [14].

2. Feichtinger's conjecture in frame theory.

3. Bourgain–Tzafriri conjecture.

All these conjectures have in fact different forms, weaker or stronger variants, etc.—a detailed account may be found in [4]. It is worth noting that up

The solution of the Kadison–Singer problem 145

to 2013 most specialists believed that they were not true, and that a counterexample would eventually be found. So it was a surprise when all these statements were simultaneously shown true by [10].

3. The method used in [10] is even stronger than described above. Actually, its first application was to a completely different problem in graph theory: the existence of certain infinite families of so-called Ramanujan graphs [9] (see also [11] for an account).

4. The most tedious proof in the above notes is that of Lemma 4.5.5. The original argument in [10] is more elegant, but uses another result of Borcea and Brändén [3] that represents real stable polynomials in two variables as determinants of certain matrices—a kind of converse to Lemma 4.5.2. The direct argument we use appears in [14].

References

[1] J. Anderson: Extensions, restrictions and representations of states on C^*-algebras, *Trans. AMS* **249** (1979), 195–217.

[2] J. Borcea, P. Brändén: Applications of stable polynomials to mixed determinants: Johnson's conjectures, unimodality, and symmetrized Fischer products, *Duke Math. Journal* **143** (2008), 205–223.

[3] J. Borcea, P. Brändén: Multivariate Pólya–Schur classification problems in the Weyl algebra, *Proc. Lond. Math. Soc.* (3) **101** (2010), 73–104.

[4] P.G. Cassazza, M. Fickus, J.C. Tremain, E. Weber: The Kadison–Singer problem in mathematics and engineering: a detailed account, *Operator Theory, Operator Algebras, and Applications*, 299–355, *Contemp. Math.* 414, AMS, 2006.

[5] P.A.M. Dirac: *The Principles of Quantum Mechanics*, Oxford University Press, 1958.

[6] R.V. Kadison, J.R. Ringrose: *Fundamentals of the Theory of Operator Algebras*, Academic Press, 1983.

[7] R.V. Kadison, I.M. Singer: Extensions of pure states, *American Jour. Math.* **81** (1959), 383–400.

[8] T. Kato: *Perturbation Theory for Linear Operators*, Springer-Verlag, 1980.

[9] A. Marcus, D.A. Spielman, N. Srivastava: Interlacing families I: Bipartite Ramanujan graphs of all degrees, *Ann. of Math.* **182** (2015), 307–325.

[10] A. Marcus, D.A. Spielman, N. Srivastava: Interlacing families II: Mixed characteristic polynomials and the Kadison–Singer problem, *Ann. of Math.* **182** (2015), 327–350.

[11] A. Marcus, D.A. Spielman, N. Srivastava: Ramanujan graphs and the solution to the Kadison–Singer problem, arxiv:1408.4421v1, to appear in *Ann. of Math.*

[12] M.A. Naimark: *Normed Algebras*, Wolters–Noordhoff, Groningen, 1972.

References

[13] C. Procesi: The invariant theory of $n \times n$ matrices, *Adv. in Math.* **19** (1976), 306–381.

[14] T. Tao: Real stable polynomials and the Kadison–Singer problem, https://terrytao.wordpress.com/tag/kadison-singer-problem/.

[15] A. Valette: Le problème de Kadison–Singer, arxiv:1409.5898v.

Index

A

Abelian Galois groups, 58
Abelian groups, 92, 109
Absolute Galois group, 58
AF algebras, 86
    triangular AF (TAF) algebras, 69
Algebraic actions and sofic entropy, 109–111
Algebraic numbers, 58
Amenability, 86–88
Amenable groups, 92
    sofic, 101
    surjunctivity, 96, 106
Amenable measure entropy (Shannon entropy), 89–95
Amenable topological entropy, 95–98
Apollo space program, 9

B

Banach algebras, 72–75
Barrier function, 131–134
Bernoulli shifts entropy classification, 83, 95
Bzout's theorem, 130
Boltzmann entropy, 89, 98–104, *See also* Sofic entropy

C

C*-algebra theory, 119
    amenability, 86–88
    C*-correspondences, 37–43
        adding tails, 45–51
        gauge-invariance uniqueness theorem, 44–45

injective correspondence, 45
interior tensor product, 41–42
C*-envelopes of arbitrary operator algebras, 56–57
C*-envelopes of operator algebras, 51–56
C*-envelopes of tensor algebras, 27–28
classification problem, 86–88
crossed product, 28
    semicrossed product, 29–31
crossed products, 65–72, *See also* Crossed product
definitions and terminology, 37–39
Kadison–Singer problem solution applications, 144, *See also* Kadison–Singer problem solution
pure states concept, 119–120
quasidiagonality, 87, 88
soficity and hyperlinearity, 88–89
twisted tensor products, 49
See also Non-selfadjoint operator algebras
Characteristic polynomials, 123, 124, 135–137
Character space of semicrossed products, 32–33
Classification of semigroups, 6
Classification problem for C*-algebras, 86–88
Classification problem for semicrossed products, 32–34
Completely contractive maps, 51–53
Complex algebraic numbers, 58

Index

Connes's embedding problem, 88
Contractive covariant
representations, 31
Crossed product, 27–28, 65–72
adding tails to
C*-correspondences, 49
classification problem, 32–34
Dirichlet or semi-Dirichlet
algebras, 68, 71
Hao–Ng isomorphism
conjecture, 72
Jacobson radical, 68–69
semicrossed product, 29–34, 39
Cuntz–Krieger–Toeplitz relations,
37, 40
Cuntz–Pimsner C*-algebra, 27, 40
adding tails to
C*-correspondences, 47–51

D

Determinant, Fuglade–Kadison, 110
Dirichlet or semi-Dirichlet algebras,
68, 71
Disc algebra, 35
Discrete valuation ring, 6, 21

E

Eigenvalues
Horn problem and sums of
Hermitian matrices, 2–5,
19–20
proof of Kadison–Singer
problem (via Paving
Conjecture), 123–124, 135
See also Kadison–Singer
problem solution
Entropy, 83–85
amenability concept, 86–88
amenable measure (Shannon
entropy), 89–95
amenable topological, 95–98
further developments, 111–112
internal and external
approximations, 85–89
Kolmogorov–Sinai approach, 89,
93–94

Ornstein's isomorphism theory,
83, 95, 104
Rokhlin entropy, 111
sofic entropy algebraic actions,
109–111
soficity concepts, 88–89
sofic measure (Boltzmann
entropy), 83–84, 89, 98–104,
107–109
sofic topological, 84, 104–107
Ergodic theory, 83, *See also* Entropy

F

Fell spectrum, 61–62
F-invariant, 111–112
Fock representation, 42–43
Flner sequence, 94, 96, 98, 101
Fuglade–Kadison determinant, 110

G

Galois group, 58
Gauge-invariance uniqueness
theorem, 30, 44–45
Graph theory
adding tails to
C*-correspondences, 46
C*-correspondence of
topological graphs, 40
Ramanujan graphs and
Kadison–Singer problem,
118, 144
reconstruction conjecture, 36–37
tensor algebra of a graph, 35–37
tensor algebra of a topological
graph, 40, 75
Grassmann manifold, 8–9

H

Hao–Ng isomorphism conjecture, 72
Hermitian matrices, sums of, 2–5,
17–19
Hilbert A-module, 37–39
Hilbert's 15th problem, 7
Horn problem, 2–5, 19–20
Hurwitz's theorem, 125
Hyperlinearity, 88–89

Index

I

Information function, 90
Injective correspondence, 38, 45
Injective von-Neumann algebras, 87, 88
Interior tensor product of C*-correspondences, 41–42
Intersection theory
    Jordan models, 21–22
    linear combinations of skeletons, 15–17
    Littlewood–Richardson rule, 11–15
    Schubert calculus, 7–10, 19
    sums of Hermitian matrices, 2

J

Jacobson radical, 68–69
Jacobson spectrum, 61–62
Jordan models, 5–7, 21–22

K

Kadison–Singer problem solution, 117–119
    Kadison–Singer conjecture, 120–121
    Kadison–Singer statement equivalents, 144–145
    overview, 123–124
    Paving Conjecture, 121–123
    Paving Conjecture proof, 123–124, 141–144
    polynomials and their roots, 123–124
        mixed characteristic polynomials, 124, 135–137
        nice families, 126–128
        real stable polynomials and barrier functions, 129–134
        univariate polynomials, 125–126
    pure states concept, 119–120
    quantum mechanics application, 144

Ramanujan graphs, 118, 144
random matrices and determinants, 123, 124, 138–141
Kolmogorov–Sinai entropy, 89, 93–94

L

Lebesgue integration, 86
Littlewood–Richardson coefficients, 9
Littlewood–Richardson rule, 11–15, 20
Local maps, 72–75

M

Matrix operations
    Jordan models, 5–7
    products, 5, 20–21
    random matrices and Kadison–Singer problem solution, 123, 124, 138–141
    sums of Hermitian matrices, 2–5, 17–19
MF algebras, 88
Mixed characteristic polynomials, 124, 135–137
Muhly–Tomforde tail of a C*-correspondence, 47
Multivariable states and non-selfadjoint operator algebras, 60–64

N

Nice families of polynomials, 126–128
Non-selfadjoint operator algebras, 27–28, 45–51
    C*-correspondences and C*-envelopes, 37–51
    crossed product, 65–72, *See also* Crossed product
    dynamics and classification, 58–64
    gauge-invariance uniqueness theorem, 30, 44–45
    Kadison–Singer problem solution applications, 144
    local maps, 72–75

motivating examples
  classification problem for
    semicrossed products, 32–34
  semicrossed product, 29–31
  tensor algebra of a graph,
    35–37
multivariable classification
  problem, 60–64
number theory interactions, 59,
  61
piecewise conjugate
  multisystems, 60
See also C*-algebra theory
Number theory
  entropy and, 83
  non-selfadjoint operator algebras
    and, 59, 61

O

Operator algebras, 27–28, 29–31
  amenability concept, 86–87
  C*-correspondences, 37–39
  C*-envelopes, 27–28, 51–57
  classification problem for
    semicrossed products, 32–34
  crossed products, 65–72, *See
    also* Crossed product
  dynamics and classification,
    58–64
  local maps, 72–75
  semicrossed product, 29–31
  tensor algebra of a graph,
    35–37
  triangular AF (TAF) algebras,
    69
  See also C*-algebra theory;
    Non-selfadjoint operator
    algebras; Tensor algebras
Ornstein's isomorphism theory, 83,
  95, 104

P

Paving Conjecture (PC), 121–123,
  141–144
Piecewise conjugate multisystems,
  60–64

Poincar dual, 10
Polynomials and their roots,
    Kadison–Singer problem
    solution, 123–124
  mixed characteristic
    polynomials, 124, 135–137
  nice families, 126–128
  real stable polynomials and
    barrier functions, 129–134
  univariate polynomials, 125–126
Pontrjagin dual, 109
Products of matrices, 5, 20–21
Pure states, 119–120

Q

Quantum mechanics application, 144
Quasidiagonality, 87, 88
Quasisimilarity, 6
Quiver algebras, 35, 40

R

Ramanujan graphs, 118, 144
Random matrices, 123, 124, 138–141
Real stable polynomials, 123,
  129–134
Reconstruction conjecture, 36–37
Rokhlin entropy, 111

S

Schubert calculus, 7–10, 19
Schubert varieties, 8–9, 14
Semicrossed product, 29–31, 39
  classification problem, 32–34
  tensor algebra of
    C*-correspondences, 39
Semi-Dirichlet algebras, 68
Semigroup classification, 6, 22
Shannon entropy (amenable measure
    entropy), 89–95
Skeleton combinations, 15–17
Sofic amenable groups, 101
  surjunctivity, 96, 106
Sofic entropy, 83–85
  algebraic actions, 109–111
  further developments, 111–112
  soficity concepts, 88–89

Index

sofic measure (Boltzmann entropy), 83, 89, 98–104
  dualizing, 107–109
sofic topological entropy, 84, 104–107
See also Entropy
Sums of Hermitian matrices, 2–5, 17–19
Surjunctivity, 96, 106

T

Takai duality, 28, 68, 71
Tensor algebras
  C*-correspondences, 39–43
  C*-envelope of, 27–28
  gauge-invariance uniqueness theorem, 45
  of graphs, 35–37

multivariable classification problem, 60–64
  of topological graphs, 40, 75
Toeplitz–Cuntz–Pimsner C*-algebra, 38–39, *See also* C*-algebra theory
Toms-Winter conjecture, 87
Topological entropy, 95–96
  amenable, 95–98
  sofic, 84, 104–107
Topological graphs, 40, 75
Torsion models, 6, 21
Triangular AF (TAF) algebras, 69

V

von Neumann algebras, 86–88, 110
von Neumann quantum mechanics, 144